North American Mixing Forum

Presentations at the 2010 AIChE Annual Meeting

Salt Lake City, Utah, USA
7-12 November 2010

ISBN: 978-1-61782-152-3

Printed from e-media with permission by:

Curran Associates, Inc.
57 Morehouse Lane
Red Hook, NY 12571

Some format issues inherent in the e-media version may also appear in this print version.

Printed by Curran Associates, Inc. (2011)

For permission requests, please contact AIChE
at the address below.

AIChE
3 Park Avenue
New York, NY 10016-5991

Phone: (203) 702-7660
Fax: (203) 775-5177

www.aiche.org

Additional copies of this publication are available from:

Curran Associates, Inc.
57 Morehouse Lane
Red Hook, NY 12571 USA
Phone: 845-758-0400
Fax: 845-758-2634
Email: curran@proceedings.com
Web: www.proceedings.com

TABLE OF CONTENTS

MIXTURE ANALYSIS OF GASES BY RAMAN IMAGING

Sascha R. Engel, Andreas Braeuer, Robert F. Hankel, and Alfred Leipertz
Lehrstuhl für Technische Thermodynamik and Erlangen Graduate School in Advanced Optical Technologies (SAOT),
Universität Erlangen-Nürnberg, Erlangen, Germany

Paper number: ID 178911

Introduction

Various processes in the field of chemical engineering are rate-determined by mixture generation and not by the subsequent reactions. The mixing degree of two different flows is represented by the two scalar quantities "mixture fraction" and/or "mixture temperature" while the mixing driving force is represented by the vectorial quantity "velocity field". Hence there is the necessity for a measurement technique that simultaneously acquires the temperature, composition and velocity field for a comprehensive characterization of the mixture formation. Furthermore this measurement technique needs to be non-intrusive in order to prevent any distortions of the flow field.

Current trends in laser imaging techniques for flow analysis point towards high speed and multi-parameter imaging. By means of the latter, it is aimed to investigate the interaction of scalar fields and vector fields during the turbulent mixing process in a multi-component system. In this context spontaneous Raman scattering is introduced as a potential optical technology suitable to achieve this goal.

Theoretical Background

The term "Ramanography" is known since 1974 when Hartley measured the distribution of one certain species molecule using planar Raman scattering [1]. Ramanography is based on the Raman Effect, which in fluids denotes, that a photon is inelastically scattered from a molecule. Consequently the scattered photon is spectrally shifted from the incident photon by the energy of the molecular transition. This so called Raman shift is species specific and its intensity is directly proportional to the number of molecules.

Ramanography was applied for the analysis of jets and turbulence in reacting and non-reacting flows under ambient pressure and in pressurized chambers at ambient temperature and under cryogenic conditions [2-6]. Due to low detection signal intensities usually methane and hydrogen are probed because of their rather high Raman scattering cross sections. In addition to multi-species investigations and local density measurements [7] the possibility of acquiring the temperature and the concentration field simultaneously was developed [8] and already used in a hydrogen internal combustion engine [9].

The composition of the jet mixing experiment is characterized by the H_2 mole fraction $x(n_{H2})$ which is defined as the number of H_2 molecules $n(H_2)$ over the total number of H_2 and N_2 molecules $n(N_2)$. As the Raman signal is directly proportional to the number of molecules, the corresponding H_2 Raman signal fraction $x(I_{H2})$ is the H_2 Raman signal intensity $I(H_2)$ over the total intensity of the H_2 and the N_2 Raman signal $I(N_2)$.

$$x(n_{H_2}) = \frac{n(H_2)}{n(H_2)+n(N_2)} \propto \bullet\, x(I_{H_2}) = \frac{I(H_2)}{I(H_2)+I(N_2)} \tag{1}$$

To correlate $x(I_{H2})$ and $x(n_{H2})$ a calibration is carried out at defined pressures by filling the injection chamber with varying H_2/N_2 compositions. The benefit of considering Raman signal ratios is that pulse-to-pulse laser energy fluctuations, inhomogeneities in the excitation light sheet, beam distortions and window fouling effects cancel out.

The temperature acquisition is based on probing the temperature sensitive distribution of molecules among their rotational energy levels. Figure 1 shows the theoretically calculated populations of several rotational J and vibrational v levels of hydrogen and nitrogen and the corresponding Raman shift in wave numbers. The population distribution across the rotational and vibrational energy levels follows the Boltzmann statistics.

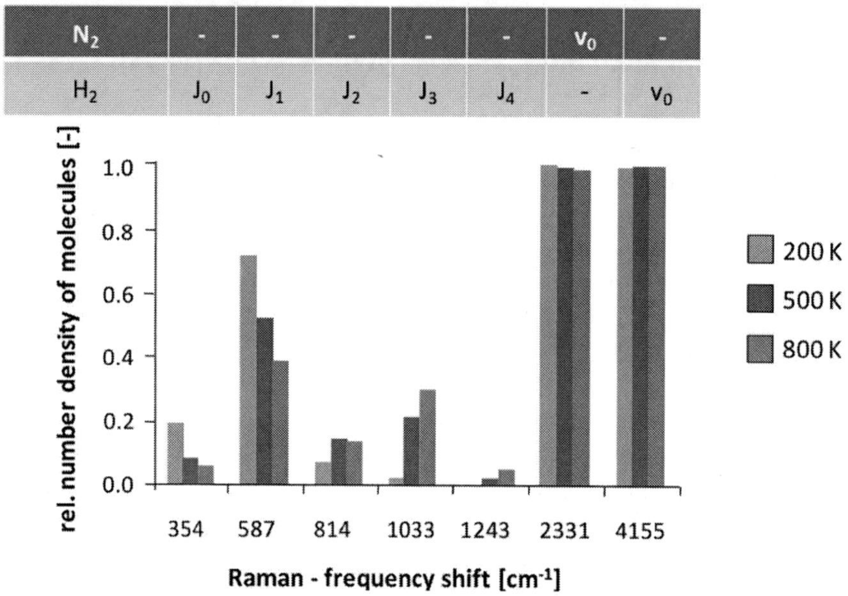

Figure 1: Population of vibrational levels (v) of hydrogen and nitrogen as well as the population of different rotational levels (J) of hydrogen for three different temperatures; according to [4].

The population of the hydrogen rotational levels within temperatures between 200 and 800 K is temperature sensitive. Contrary, the population of the nitrogen and hydrogen vibrational levels is almost constant in the considered temperature range. Hence the number of species i molecules in the vibrational ground state v_0 is equal to the total number of species i molecules present between 200 K and 800 K,

$$N(i)_{v_0} = N(i)_{total} \tag{2}$$

which is true for H_2 and for N_2. Thus the temperature can be obtained by measuring the signal intensities of the hydrogen rotational levels. High temperature resolution can be achieved, if the population ratio of J_3 and J_1 is measured. Other combinations are possible, but both energy levels J_1 and J_3 are sufficiently populated in the temperature range of interest.

The combination of Ramanography and particle image velocimetry (PIV) allows the simultaneous acquisition of scalar and vectorial data. PIV is an optical measurement technique where a fluid is seeded with a trace of particles small enough to follow the gas flow without disruption. The flow is illuminated subsequently by two laser pulses. The incident light is scattered elastically from the particle and is termed Mie scattering here. The Mie scattering of both incident laser pulses is detected one after the other by a CCD-camera which can take two images synchronized to the time delay between the incident laser pulses. A cross correlation algorithm computes the shift of the scattering patterns in both images and represents this shift as a vector field.

Setup

Figure 2 shows the basic setup for the listed applications in this paper. As an **excitation** source a laser cluster consisting of four pulsed Nd:YAG lasers with a beam diameter of 8 mm is used. Each s-polarized laser is operated at 532 nm, with a repetition rate of 10 Hz and a pulse duration of 10 ns (full width half maximum). Exposure times, determined only by the excitation timing, are sufficiently low to freeze turbulence. By the use of different mirrors and lenses a vertical light sheet of 20 mm height and 200 µm thickness is launched into the injection chamber through the center of the H_2 jet.

Hydrogen is injected into a pressurized chamber at room temperature. **The measuring chamber** is continuously purged with nitrogen. Flow rates and initial temperatures are known for both hydrogen and nitrogen. Further the pressure in the chamber is defined. For PIV measurements droplets are seeded in the hydrogen flow.

The **detection** system is equipped with a signal collection lens at a distance of exactly f from the laser light sheet plane for two reasons. First, the desired signals are detected through a wide solid detection angle. This results in high signal intensities. Second, the signals are collimated behind the signal collection lens. This results in very efficient signal separation and signal purification as dielectric beam splitters and filters show their best performance for collimated radiation. Three cameras, representing the camera cluster, are equipped each with an additional lens (Nikon Nikkor 50 mm; f/1.2). With this configuration an area of 23 µm x 23 µm is imaged onto one single pixel (not binned) of the CCD chip. The CCD chip of the Raman cameras is binned 8 x 8 resulting in a spatial resolution of 184 µm x 184 µm, while the CCD chip of the PIV camera is not binned. Dichroic mirrors direct the detection signal to the specified camera. Different filters avoid the crosstalk of the detection signals.

The experimental setup is customized for the presented investigations. Therefore timing of the excitation, the filter setup and the dichroic mirrors are changed. Detailed information can be found elsewhere [8, 10].

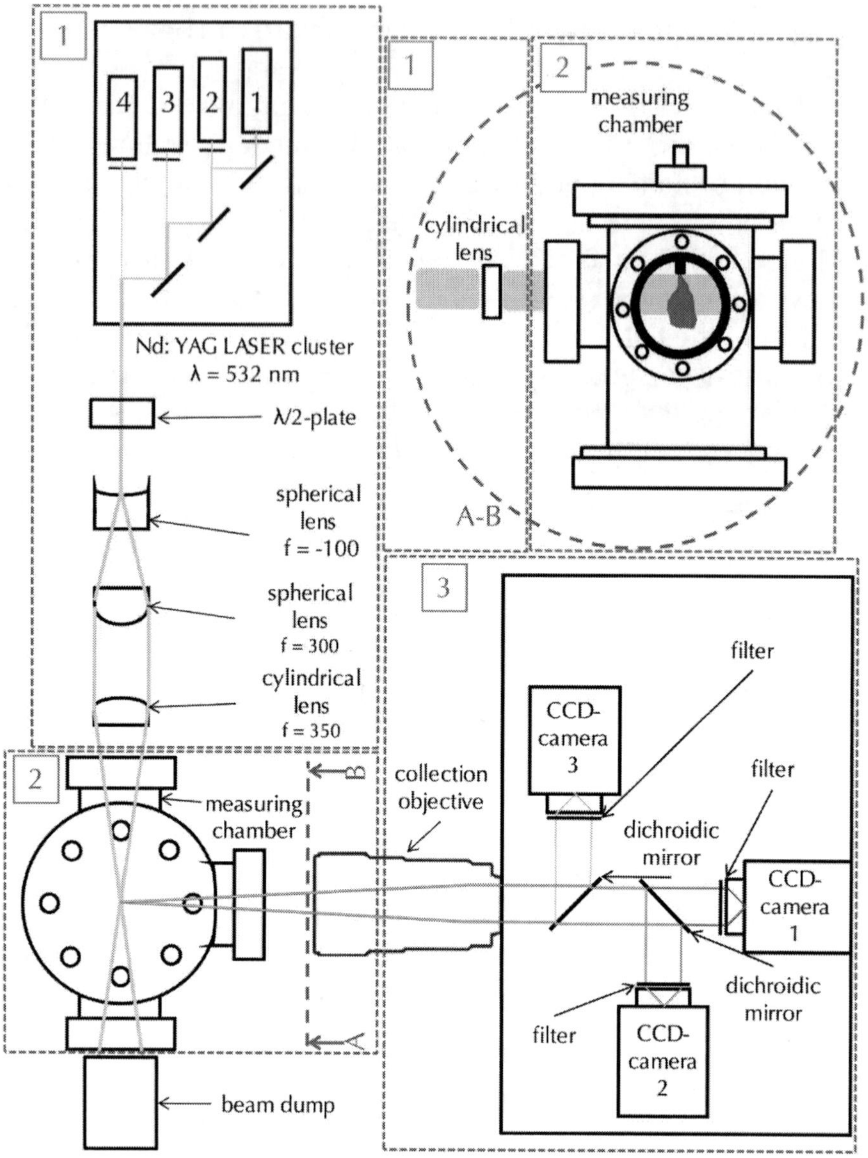

Figure 2: Basic scheme of the setup for the listed applications. [1] Excitation; [2] measuring chamber; [3] detection.

Results

The mixture formation of hydrogen and nitrogen was investigated regarding the **temperature and concentration field** by Ramanography [8]. The Raman process was excited with a Nd:YAG Laser at 532 nm. Hence the relevant Raman signals result at 549 nm (J_1 rotational level H_2), 564 nm (J_3 rotational level H_2) and 607 nm (v_0 vibrational level N_2). These three signals were separated by color selective (dichroitic) mirrors and detected by three CCD-cameras. Figure 3 shows the simultaneously acquired concentration and temperature field of a hydrogen jet in a nitrogen environment.

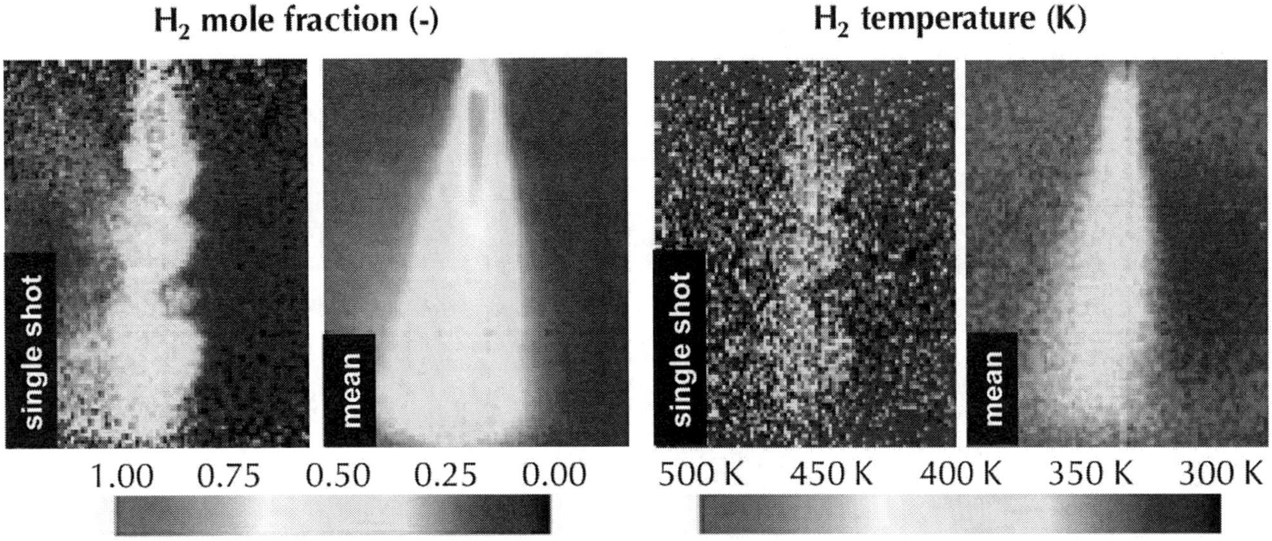

Figure 3: Mean and single-shot hydrogen mole fraction $x(n_{H2})$ and hydrogen temperature $T(H_2)$ images of a hot (473 K) hydrogen jet which is injected into nitrogen at room temperature (297 K) at an absolute pressure of 1.1 MPa.

This measuring technique was combined with PIV to receive in addition to the scalar field also the **vector field**. As Ramanography is species selective the signals of different species are spectrally displaced from the incident laser pulse. Supplementary signal spectra are not overlapping for small molecules (typically gases). Accordingly Raman scattering is perfectly suited for the simultaneous application with other optical measurement techniques in multi parameter experiments. One disadvantage of spontaneous Raman scattering suffers from very small Raman scattering cross sections, which are several orders of magnitude smaller than Mie scattering cross sections from droplets or particles. Consequently for combining spontaneous Raman scattering with droplet seeded PIV experiments, the detection of pure and reliable Raman signals in the gas phase, without interference from Mie signals, is most challenging. The signal separation is achieved by an ascertain set of filters and dichroitic mirrors. Figure 4 shows the combination of a PIV vector field with a Ramanography composition field for a droplet seeded H_2 flow injected in a purged nitrogen chamber at 0.95 MPa [10].

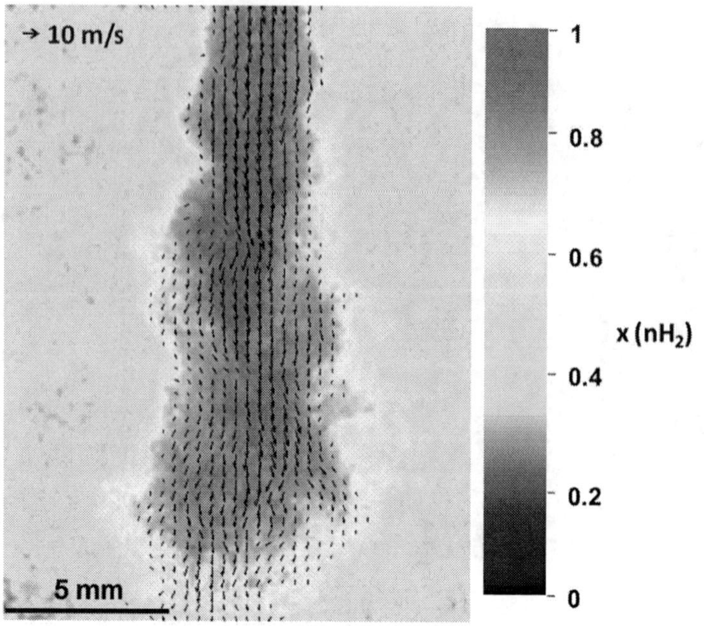

Figure 4: Combination of the PIV vector field and the Ramanography composition field when a droplet seeded H_2 flow is injected into a N_2 purged chamber at 0.95 MPa and at room temperature (spurious vectors are removed).

Moreover the limits of Ramanography regarding the temperature range were probed by injecting **hydrogen at low temperatures** in a purged nitrogen chamber. The temperature was calculated using the adiabatic mixture temperature. Figure 5 shows the temperature and mole fraction distribution of the injection of hydrogen in a purged nitrogen chamber. The spatial resolution of the presented images in Figure 4 and Figure 5 is 184 µm x 184 µm. The cloud-shaped propagation of the hydrogen jet is owed to the usage of a hydrogen hollow cone injector.

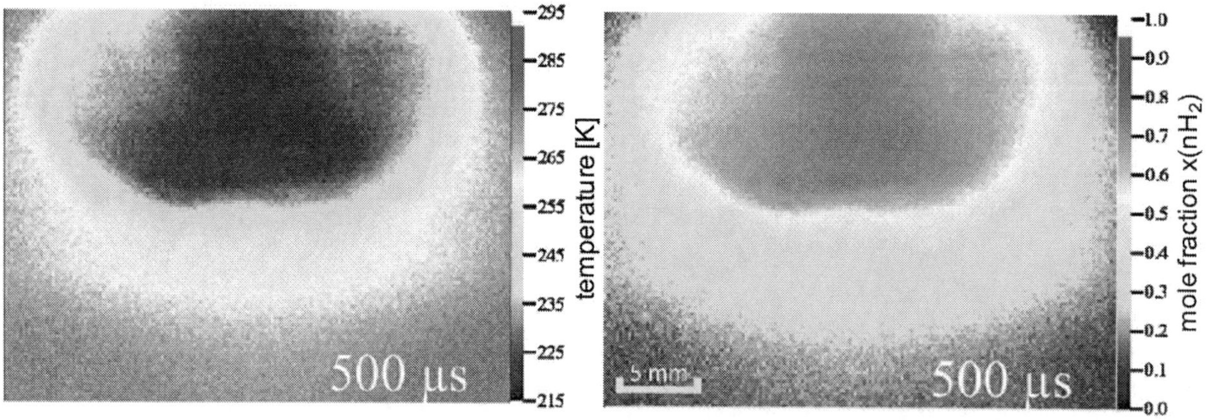

Figure 5: Temperature and mole fraction distribution of the injection of hydrogen in a purged nitrogen chamber at 0.4 MPa and at injection temperatures of 200 K. The image was taken 500 µs after start of injection.

Conclusion

By use of the inherent properties of Raman scattering like species selectivity, direct quantifiability and independency from tracers, a powerful and flexible laser diagnostic technique provides deep insight in mixing processes in the gas phase. Several parameters essential for the mixture formation are measured simultaneously. Short laser pulse durations facilitate the resolution of turbulent processes. Finally the experimental data can be used to validate numerical models. Further improvements of the measuring strategy as well as ongoing development of laser systems and cameras will extend the application range steadily.

The authors gratefully acknowledge funding of parts of this work by the Bavarian Research Association (BFS) and the German Research Foundation (DFG), which also funds the Erlangen Graduate School in Advanced Optical Technologies (SAOT) in the framework of the German excellence initiative. The authors also acknowledge equipment support of BMW Group.

References

1. D. L. Hartley, *Laser Raman gas diagnostics* (Plenum, 1974), pp. 1151-1157.
2. M. B. Long, P. S. Levin, and F. D.C., "Simultaneous two-dimensional mapping of species concentration and temperature in turbulent flames," Opt. Lett. **10**, 267-269 (1985).
3. D. C. Kyritsis, F. G. Felton, and F. V. Bracco, "Instantaneous, two-dimensional, spontaneous Raman measurements of hydrogen number density in a laminar jet using an intra-cavity configuration," Int. J. Altern. Propul. **1**,(2007).
4. A. R. Masri, R. W. Dibble, and R. S. Barlow, "Raman-Rayleigh scattering measurements in reacting and non-reacting dilute two-phase flows," J. Raman Spectrosc. **24**, 83-89 (1993).
5. R. D. Woodward and D. G. Talley, "Raman imaging of transcritical cryogenic propellants," in paper No. AIAA 96-0468 (American Institute of Aeronautics and Astronautics, 1996),
6. D. F. Marran, J. H. Frank, M. B. Long, S. H. Stårner, and R. W. Bilger, "Intracavity technique for improved Raman/ Rayleigh imaging in flames," Opt. Lett. **20**, 791-793 (1995).
7. S. Dowy, A. Braeuer, K. Reinhold-López, and A. Leipertz, "Laser analyses of mixture formation and the influence of solute on particle precipitation in the SAS process," The Journal of Supercritical Fluids **50**, 265-275 (2009).
8. A. Braeuer and A. Leipertz, "Two-dimensional Raman mole-fraction and temperature measurements for hydrogen-nitrogen mixture analysis," Appl. Opt. **48**, B57-B64 (2009).
9. S. Engel, A. Braeuer, P. Koch, and A. Leipertz, "Ramanography for the H2-mixture Formation Analysis under Engines Relevant Conditions," in *Engine Combustion Processes*, (Berichte zur Energie- und Verfahrenstechnik, 2009),
10. A. Braeuer, S. R. Engel, R. F. Hankel, and A. Leipertz, "Gas mixing analysis by simultaneous Raman imaging and particle image velocimetry," Opt. Lett. **34**, 3122-3124 (2009).

(35b) Mixing Viscous Fluids Using a Reconfigurable Split-and-Recombine Static Mixer

Emilio J. Tozzi[1], David M. Lavenson[1], Michael J. McCarthy[2] and Robert L. Powell[1], (1)Chemical Engineering and Materials Science, University of California Davis, Davis, CA, (2)Biological and Agricultural Engineering, University of California Davis, Davis, CA

Mixing of highly viscous materials at low Reynolds numbers can be achieved by stretching and folding motions. Pure shear flow can achieve a stretching motion that grows approximately linearly with time. Extensional flows, on the other hand, are potentially more effective due to the exponential stretching in time of fluid elements. Split and recombine (SAR) static mixers employ extensional flow by rearranging fluid into stacks of layers which duplicate after each mixing stage. The use of a SAR mixer is proposed for homogeneous dispersion of small amounts of material into a highly viscous fluid. A prototype was constructed by stacking a series of thin channel cross sections, which are easily constructed from sheets of material and can be rearranged for optimization and design purposes. We discuss the characterization of the mixer performance regarding flow patterns, pressure drop and mixing quality via numerical simulations and magnetic resonance imaging experiments.

See more of this Session: Mixing in Single Phase Systems
See more of this Group/Topical: North American Mixing Forum

(35d) Experimental Determination, Computational Prediction, and Modeling-Based Analysis of Mixing Time in a Fully Baffled Stirred Vessel Equipped with a High Shear Homogenizer

Micaela Caramellino, Otto H. York Dept. of Chemical, Biological and Pharmaceutical Engineering, New Jersey Institute of Technology, Newark, NJ and Piero M. Armenante, Otto H. York Department of Chemical, Biological and Pharmaceutical Engineering, New Jersey Institute of Technology, Newark, NJ

Shear sensitive batch processes are commonly found in the chemical and pharmaceutical industries. These processes can be enhanced by introducing a high shear homogenizer in a stirred tank system agitated by a conventional primary impeller. This arrangement combines the blending/suspending ability of the primary impeller with the shear effects introduced by the homogenizer in a single device. In this work, the mixing time required to homogenize the system's liquid content in a fully baffled, stirred tank equipped with a centrally mounted low speed primary impeller (disk or pitched blade turbine) and an additional high-shear rotor-stator homogenizer was experimentally determined using a colorimetric technique coupled with image analysis. The effect of the presence of the homogenizer on mixing time was assessed for different operating variables, including tank size, type of primary impeller and the agitation speeds of both the homogenizer and the low speed impeller. The experimental results were compared with predictions based on two different computational/mathematical modeling approaches. The first was a CFD-based approach where mixing time was predicted by computationally determining the rate of dispersion of a tracer. This approach utilized CFD simulations to predict the velocity distribution, which were also experimentally validated. The second approach was based on the development of a compartmental model in which the flow between compartments was assumed to be directly proportional to the pumping capacity of all the mixing devices in the system. Experiments data show that the overall blend time produced by the primary impeller alone was significantly reduced by the presence of the high shear homogenizer in the tank. The same trend was observed at different primary impeller speeds, homogenizer speeds, and total tank volumes. Good agreement between the experimentally measured mixing times and the predictions of both the CFD-based model and the compartmental model was found.

See more of this Session: Mixing in Single Phase Systems
See more of this Group/Topical: North American Mixing Forum

(35e) Active Mixing Enhancement in Confined Mixing Layer

Wei Zhao and Guiren Wang, Mechanical Engineering, University of South Carolina, Columbia, SC

Drastic mixing is achieved in confined mixing layer and wake flow by actuating at a narrow frequency band, i.e. optimal frequency, in a wide range of Re. The potential causes, such as hydroacoustic resonance mechanism, fluid-solid interactions, the inherent frequency of actuating system and sloshing wave, are in detail investigated by parametric study using flow visualization. By changing the parameters, such as the length and diameter of mixing chamber, the length of settling chamber and the fixing methods of all the parts of the water tunnel system, the optimal frequency, i.e. 5.3 Hz in our setup is not changed. The mixing effect is not apparently affected either. Further investigation indicates the shape of trailing edge of splitter plate strongly affect the mixing effect. The sharper the trailing edge, the stronger the mixing. The flow field without the splitter plate is also tested. It is found that in this case there is no apparent difference with or without the forcing. Hence, it is obvious that the splitter plate, i.e. the specific configuration of flow field plays a key role in the mixing enhancement process. Confined mixing layer is another key factor. It is believed that the confined region induces the rapid growth of 3-dimensional flow fluctuations and vortex structures, which lead to the rapid mixing. The nature of the unforced and forced flow field, the receptivity, the spectrum characteristics and the vortex structures are all under investigation by PIV.

See more of this Session: Mixing in Single Phase Systems
See more of this Group/Topical: North American Mixing Forum

(35f) Determination of the Local Micromixing Structure in Laminar Flows

Edward P. L. Roberts, School of Chemical Engineering and Analytical Science, University of Manchester, Manchester, United Kingdom

In rapidly mixing flows with low diffusion rates and/or fast reactions, accurate calculation of the local concentration distribution becomes impractical. One approach is to consider mixing segregated fluids and follow the interface between the two fluids as it stretches in the flow. For a rapid mixing flow such as a chaotic or turbulent flow, the area of the interface between segregated fluids grows exponentially, so that determination of the location of the interface rapidly becomes unfeasible. In this study, a method for calculating the local lamellar structure (i.e. the orientation and thicknesses of the lamellae) for any advection time and at any location in the flow is developed for two dimensional flows. Locally the lamellar structure is assumed to be one dimensional, and the orientation of the lamellae can be determined from the direction corresponding to the maximum stretching rate for fluid arriving at the location. The lamellar structure can be readily determined by mapping a short finite line, oriented perpendicular to the orientation of the lamellae, backwards in time to the initial conditions. Once the line is returned to its initial location the intersection of the line with the initial interface between the segregated fluids can be used to accurately determine the detailed lamellar structure, in particular the thicknesses of the lamellae. The accuracy of the method is demonstrated for two deterministic chaotic flows, the sine flow and a piecewise linear flow, the saw-tooth flow. The use of the piecewise linear flow enables accurate testing of the method as precise simulations of the full interface can be obtained for relatively long advection times, where the initial interface is a polygon.

See more of this Session: Mixing in Single Phase Systems
See more of this Group/Topical: North American Mixing Forum

(107a) Minimum Agitation Speed for Solid Suspension in a Torispherical-Bottomed Pharmaceutical Mixing Tank Agitated with a Retreat-Blade Impeller Under Different Baffling Conditions

Dilanji Wijayasekara and Piero M. Armenante, Otto H. York Department of Chemical, Biological and Pharmaceutical Engineering, New Jersey Institute of Technology, Newark, NJ

The pharmaceutical and specialty chemicals industries often utilize glass-lined reactors in the manufacturing of Active Pharmaceutical Ingredients (API) because of their material compatibility with most reactants. These reactors are typically equipped with a retreat-blade impeller placed close to the bottom of the tank a single baffle mounted from the top. The mixing performance of such reactors has not received significant attention in the literature, despite their ubiquitous presence in the pharmaceutical industry. In this work the minimum agitation speed, Njs, required to just suspend solid particles dispersed in water was experimentally determined for a 60 L, scale-down version of a glass-lined pharmaceutical vessel provided with a torispherical bottom and agitated with a retreat-blade impeller under different baffling configurations, i.e., unbaffled, partially baffled, and fully baffled tank. Njs for the same tank but equipped with a different agitation system, namely an axial impeller, was also experimentally determined for the purpose of comparing of performances of the two systems. The effect of impeller off-bottom clearance and the vessels liquid level on the minimum agitation speed were also experimentally studied. Njs was experimentally determined using Zwietering's method, requiring that the solids do not rest on the tank bottom for more than 1-2 seconds. The value of Njs was found to depended strongly on the type of baffling, and was highest in the unbaffled tank and with solid particles 150μm in size and lowest in the partial baffled system.

See more of this Session: Mixing in Multi-Phase Systems I
See more of this Group/Topical: North American Mixing Forum

(107b) Pulse Jet Mixers - the Known and Unknown about Solids Suspension

David S. Dickey, MixTech Inc., Dayton, OH

Pulse Jet Mixers (PJMs) are unique for large-vessel mixing equipment because they have no moving parts. PJMs are controlled and operated entirely by air-driven fluidic devices. Besides having unique mechanical features, PJMs also have a unique cyclical operation, alternating between a power jet period and a suction recovery period. This on and off operation creates unique results for all aspects of mixing, but especially for solids suspension. The actual suspension of solids occurs almost entirely in a start-up mode as rapidly-settling solids are lifted from the vessel bottom during the power jet period of the cycle. During the suction recovery period, the same rapidly settling particles may return to the bottom of the vessel, awaiting resuspension during the next power pulse. So each power pulse involves the initial and dynamic lifting of settled particles. In large-scale vessels, the power jet period could be roughly a minute long, while the suction recovery period could last for three or four minutes.

Extensive testing in scale-model systems has provided interesting data about how pulsed suspension of solids both works and fails. During the power pulse, a range of particle sizes and densities have been studied to decide the pulse requirements for successful suspension. Simultaneously, the lengthy suction recovery period virtually assures that a "uniform" suspension can never be achieved for rapidly settling particles. While more than two-thousand test results have looked at scale-up and operating effects in three vessel sizes, 0.367 m, 0.860 m, and 1.778 m diameters, uncertainties about some variable effects remain. Additional geometrically-similar prototypic tests have been conducted in a 1.10 m diameter vessel.

As more testing is done, the effects and interactions between different size and density particles continue to make any analysis or evaluation more difficult. While fine, low-density particles will remain suspended between pulses, their presence can affect the suspension of the rapidly settling particles. The studies have focused on the suspension of small quantities of rapidly settling particles. The rapidly settling particles may be either moderate size high density particles or large diameter moderate density particles. High density or large size has different effects on solids suspension, even for particles with the same terminal settling velocity. Other factors, such as cohesive effects within the solids also influence suspension results.

The literature on mechanical mixing contains only limited information about the effects of variables like liquid level, bottom shape, solids concentration, vertical distribution of solids, initial and resuspension of settled solids, bimodal and broad particle size distributions, but especially the scale-up of these different variables. Dynamic operation for solids suspension with PJMs provides some insights into the operational and mechanistic characteristics of all types of solids suspension, including many opportunities for further investigation. The primary objective of this study is to prevent accumulation of solids in any vessel as they are filled and partially emptied in a batch-wise operation.

See more of this Session: Mixing in Multi-Phase Systems I
See more of this Group/Topical: North American Mixing Forum

(107c) MIXING ANALYSIS of Neutrally Buoyant PARTICLES of Finite SIZE IN COMPLEX FLOW

Alison Hirsch[1], Boyang Zhang[1], Che-Ying Kuo[2] and **Hang Lu**[1], (1)School of Chemical and Biomolecular Engineering, Georgia Institute of Technology, Atlanta, GA, (2)Chemical & Biomolecular Eng., Georgia Institute of Technology, Atlanta, GA

Microfluidic devices is widely used for chemical and biological applications. Chaotic mixing using the staggered herringbone (SHB) mixer, for example, is a common strategy for effective mixing of solutes. In mixing particles for neutrally buoyant cells of finite size, however, the distribution is dissimilar to that of the solutes and not uniform after long mixing time. To better understand the physics and evaluate particle mixing strategies in dilute suspensions, we developed both experimental and modeling techniques. We use a simple epifluorescent technique - Single-field Three-dimensional Epifluorescent Particle (STEP) imaging - which can find the particle depth (z) from the fluorescence intensity patterns of the moving particles and generate 3D particle distributions. The exact location of individual particles can be recognized and particle tracking experiments can also be performed easily and less expensively than confocal microscopy. We found that particle and solute distributions after a number of cycles in a slant groove mixer (SGM) under the same flow conditions are dissimilar. We found that many factors affect the particle distribution: flow Re, particle size and particle Re, and groove geometry. Mixing patterns are particularly sensitive to the particle size and groove geometry within the Re number range examined. We hypothesize that the finite particles have a defined set of accessible streamlines; streamlines that come within a distance equal to the particle's radius to the wall cannot contain particles. Preliminary computational fluid dynamics simulations support this hypothesis. The experimentation and modeling tools allow us to further understand the detailed mechanisms behind this phenomenon and will allow us to efficiently design microfluidic unit operations (mixing, separation and sample concentration) on particle or cell suspension of finite size.

See more of this Session: Mixing in Multi-Phase Systems I
See more of this Group/Topical: North American Mixing Forum

(107d) Pilot-Scale Testing of Solid-Liquid Mixing for Radioactive Waste Treatment

Michael Poirier and Zafar Qureshi, Savannah River National Laboratory, Aiken, SC

The Savannah River Site (SRS) is developing processes to treat radioactive liquid waste in million gallon waste tanks. In one of these processes, salt solution containing entrained solids is added to a waste tank. The process then adds monosodium titanate (MST) to the process to sorb strontium and select actinides. The tank is mixed with rotating jet pumps. Following the sorption process, the salt solution is filtered to remove the solid particles and processed through an ion exchange process to remove cesium.

The authors designed and constructed a nominal 1/10 scale tank to evaluate the process and size the pumps. The testing will determine the pump flow rate to mix the MST, so the sorption process is complete, the mixing time required for sorption, and the pump flow rate needed to suspend solid particles that settle in the tank for extended durations at elevated temperatures.

The authors will discuss the test results.

See more of this Session: Mixing in Multi-Phase Systems I
See more of this Group/Topical: North American Mixing Forum

(107e) Dispersion of Water in Viscous Oils in a High Shear Mixer

Paul E. Rueger and Richard V. Calabrese, Chemical & Biomolecular Engineering, University of Maryland, College Park, MD

The dispersion of two immiscible fluids and the resulting drop size distribution has been a subject of study for quite some time. For emulsions and fine dispersions, it has been found that using rotor-stator mixers for this purpose is superior to the employment of conventional stirred tanks because of the high energy dissipation density near and within the rotor-stator mill head. Most of the studies performed in this area have been done with a low viscosity continuous phase and a viscous drop phase. This is problematic because there is little information available for the dispersion of low viscosity liquids into high-viscosity continuous phase systems. As a result, it has not been possible to correlate the dispersed drop size distribution (DSD) with the disruptive and cohesive forces acting on the drops in these systems. Such information finds application in the scale-up of numerous dispersion processes with a high viscosity phase such as the formation of water droplets in crude oil. This study investigates the effect on the DSD of a water-in-oil dispersion in a Silverson L4R batch mixer with changing agitation rate, wettability of high-shear mixer surfaces, dispersed phase fraction, presence of salt in the dispersed phase, and continuous phase viscosity. The interfacial tension between the oil – Crystal Oil 500FG – and water is also modified by the addition of an oil-soluble surfactant, Tergitol NP-4. The wettability of the surfaces of the mixer that cause high shear rates – the ones that may influence the resulting drop size – is modified by treatment with dimethyldichlorosilane to make these stainless steel surfaces more hydrophobic. The continuous phase viscosity is controlled with temperature since the oil's viscosity is a strong function of temperature. The results of these experiments are used to postulate correlations for drop-size with the pertinent dimensionless number(s). Furthermore, they are used to investigate the relative importance of the drop size changing mechanisms of drop breakup and coalescence as a result of phase fraction and the other parameters.

See more of this Session: Mixing in Multi-Phase Systems I
See more of this Group/Topical: North American Mixing Forum

(107f) Investigation of Mass Transfer in Poly-Disperse Gas-Liquid Systems by Using a Multi-Variate Population Balance and CFD

Antonio Buffo, **Daniele Marchisio** and Marco Vanni, DISMIC - Politecnico di Torino, Torino, Italy

The simulation of momentum and mass transfer together with chemical reactions is complicated by the fact that the gas phase is poly-dispersed, namely it is constituted by bubbles characterized by a distribution of velocity, size and composition values. Phase coupling can be successfully described only if the modeling approach acknowledges the existence of this distribution, whose evolution in space and time is governed by a multi-variate population balance equation. A very efficient is represented by quadrature-based moment methods, where the evolution of the relevant properties is decoupled by tracking pure moments only and where the so-called closure problem is overcome by employing a quadrature approximation. When for example a quadrature approximation with two nodes is used the bubbles are represented by two distinct bubble classes, each characterized by its own velocity, size and composition. This procedure allows to accommodate four moments of the distribution with respect to bubble size (mean, variance and skewness) and two moments with respect to bubble velocity and composition (mean and variance). Another interesting advantage is that the final equations can be easily implemented in commercial Computational Fluid Dynamics codes. The method is here applied to the simulation of a system investigated by other authors for which experimental data are available in the literature for model validation. In these works mass-transfer was studied in conjunction with bubble coalescence and breakup. As it possible to see in the figure, reporting the contour plot of characteristic size of bubble classes 1 and 2 and mean Sauter diameter (m), bubbles enter the reactor with a mean size of about 3 mm completely determined by the gas sparger and then due to coalescence and breakage the bubble size distribution completely changes, deeply affecting phase coupling. Simple mass transfer experiments are here used for model validation: the reactor is fluxed with nitrogen and then after switching to oxygen the time evolution of species concentration in the liquid and gas phase is tracked. Model predictions and comparison with experimental data show that in order to properly describe the global mass transfer rate it is essential to represent the poly-dispersity of the gas phase, proving the validity of the approach.

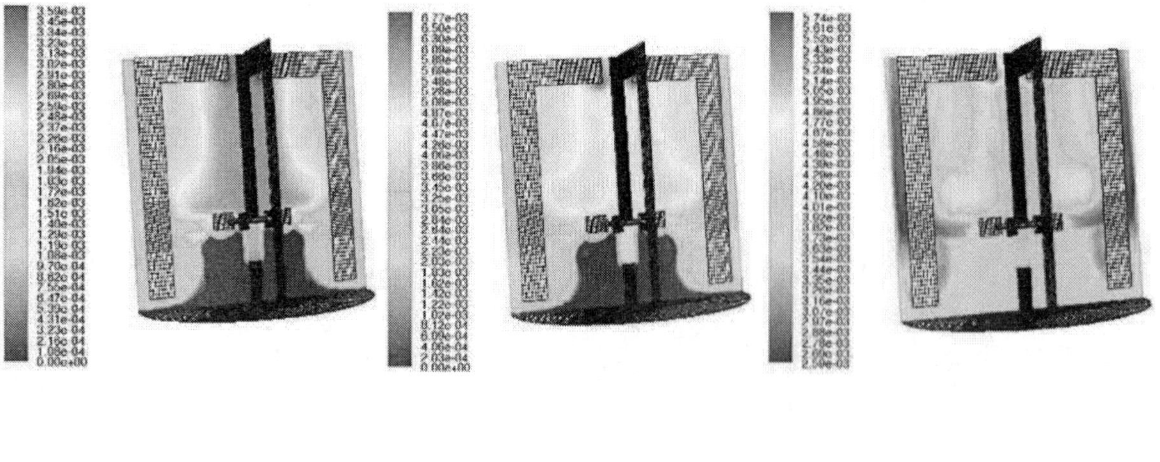

(158a) Solids Suspension in Tall Tanks

Richard F. Cope[1], **Madan Somasi**[2] and Kishore K. Kar[1], (1)The Dow Chemical Co., Midland, MI, (2) The Dow Chemical Company, Midland, MI

Uniformly suspending slurry solids in process vessels (reactors, crystallizers, storage tanks, etc.) of various sizes and shapes is a significant challenge in the chemical industry. This may be a bigger issue in manufacturing processes using glass-lined vessels, such as suspension polymerization, where large capacity vessels are disproportionately tall due to size constraints on the vessel manufacturer's glass-firing furnace. While tall tanks offer the enticement of higher capacity with a lesser foot print, uniformly suspending solids within such vessels is still a very daunting challenge.

The extensive work in solids-liquid systems is well represented in Paul et al. (2004) and such recent studies as (Brown et al., 2007; Himmelsbach et al., 2007; Spogis and Nunhez, 2007; Khazam and Kresta, 2006; Ditl and Rieger, 2006; Bao et al., 2005). These works involve experimentation and/or computer modelling to examine a variety of floating or settling solids, and traditional and/or new impeller types operating at various conditions of agitation speed and flow direction. The research described in this document seeks to understand the performance of a multi-tier agitation system consisting of either axial flow hydrofoils or mixed flow pitched blade turbines in uniformly suspending a specific settling solid in water throughout a "tall" laboratory vessel.

This study determined how the solids distribution changes as a function of solids loading (10, 20, 30, and 40 wt%), mixing rate (100, 150, 200, 250 RPM), and impeller type (pitched blade turbine (PBT) or wide-blade hydrofoil (Lightnin A320 or A340)). All work was conducted in a 17.5 in. ID Plexiglas® tank, with a slurry height of 35 in. At fixed operating conditions, the solids concentrations were obtained at four different axial tank locations. The results indicate that both impeller types produced a uniform solids suspension when the rotation rate produced an energy input of at least 6 HP/1000 gallons for the hydrofoils, and 7 HP/1000 gallons for the PBTs. Significantly higher energy or rotation rates improved the solids distribution only slightly because, at some point, flow within the vessel appeared to become choked.

See more of this Session: Mixing in Multi-Phase Systems II
See more of this Group/Topical: North American Mixing Forum

(158b) Single and Multiphase Mixing in Partially Filled Stirred Vessels Using Computational Fluid Dynamics and Particle Image Velocimetry

Shilan Motamedvaziri, Otto H. York Department of Chemical, Biological and Pharmaceutical Engineering, New Jersey Institute of Technology, Newark, NJ and Piero Armenante, Chemical, Biological and Pharmaceutical Engineering, New Jersey Institute of Technology, Newark, NJ

In many industrial applications, mixing vessels have liquid height-to-tank diameter ratio H/T equal to, or larger than, 1. However, there are many instances where this ratio is lower than 1, as in all those cases in which the vessel is emptied. Even when H/T<1, sufficient agitation must still be provided in order to attain the process objectives. When the impeller submergence is reduced as a result of lowering the liquid level, the fluid dynamics of even a single-phase stirred liquid can become quite complex, with different regimes possibly existing depending on the geometric characteristics of the system (such as impeller clearance, liquid height, or liquid head above the impeller). The objective of this work is to determine the minimum liquid level and the critical impeller submergence for a standard impeller off-bottom distance (C/T=1/3) where adequate mixing process can still be achieved both in a single liquid phase and solid-liquid suspension.

Computational Fluid Dynamic (CFD) simulation as well as Particle Image Velocimetry (PIV) was used here to study the velocity profile in flat-bottom baffled vessels equipped with a single Disk Turbine (DT) for different height-to-tank diameter ratios when the liquid level is smaller than the tank diameter. A commercial pre-CFD mesh generator (Gambit 2.4.6) coupled with a CFD software package (Fluent 6.3.26) was used. A standard k-ε model as well as a realizable k-ε model coupled with enhanced wall treatment was used to model the turbulence flow. Results were obtained using a multiple references of frames (MRF) approach. Transient simulations based on a sliding mesh technique were additionally conducted. Experimental work was also carried out with a single liquid phase as well as in solid-liquid suspensions with 0.5% dispersed solids.

In general, good agreement between the experimental data and the predicted results for the velocities distribution, power number, and pumping number were obtained. Both the experimental and the computational results show that there is a minimum liquid level (or minimum impeller submergence) below which: (1) the macroscopic flow pattern changes substantially; (2) the power number and the pumping number drop significantly; and (3) air entrainment occurs and solids suspension becomes problematic.

See more of this Session: Mixing in Multi-Phase Systems II
See more of this Group/Topical: North American Mixing Forum

(158c) The Effect of Particle Properties and Solids Loading On Suspension of Mixed Solids

Inci Ayranci and Suzanne Kresta, Chemical and Materials Engineering, University of Alberta, Edmonton, AB, Canada

Solids suspension is a complicated operation with many variables related to solid and liquid properties and impeller and tank geometry. The objective of most solid-liquid mixing operations is to utilize the entire surface area of the solids for mass transfer purposes. This requires operating at N_{js}, just suspended speed, where complete off bottom suspension condition is satisfied. N_{js} increases with increasing solid density and particle size. It is relatively easier to determine N_{js} and design when the slurry is composed of unimodal solids; however, this is hardly ever the case in industry. The slurries in industrial operations, such as mineral processing, are composed of many solid phases with varying density and particle sizes. The design correlations currently available in the literature are for unimodal slurries; therefore, it is harder to design for mixed slurries. In this study we aim to investigate the behaviour of the suspension of slurries with mixed solids at varying solids loadings. For this reason both unimodal and mixed slurry N_{js} data were observed. The mixed slurries were composed of two solid phases at specific solids loadings for each phase. The unimodal slurries were composed of each of these solid phases at the specific solids loadings in mixed slurry experiments. This allows us to predict the effect of presence of a second solid phase on N_{js}. Glass beads, bronze powder, ion exchange resin and nickel were used in the experiments. All these particles differ in particle size and density. The total solids loadings were varied from 1.5 to 55 wt% (weight percent). Two impellers were tested over a range of off-bottom clearances: Lightnin A310 and down pumping 45° PBT. The results showed that the dense particles dominate at low solids loadings; therefore, N_{js} is determined by the dense particles. At high solids loadings the particle-particle interactions become significant and a different trend is seen in the results.

See more of this Session: Mixing in Multi-Phase Systems II
See more of this Group/Topical: North American Mixing Forum

(158d) Modeling of Nucleation and Transport of Particles in Pharmaceutical Multiphase Reactors

Michael C. Gruber[1], Stefan Radl[1], Andreas Eitzlmayr[1], Daniele Suzzi[2], Christina Petschacher[3], Andreas Zimmer[3] and Johannes Khinast[1], (1)Institute for Process and Particle Engineering, Graz University of Technology, Graz, Austria, (2)Research Center Pharmaceutical Engineering (RCPE), Graz, Austria, (3)Department of Pharmaceutical Technology, University of Graz, Graz, Austria

Nucleation and transport of particles in multiphase reactors are two important mechanisms that significantly influence the final product properties. Modeling these mechanisms is, however, extremely challenging and different approaches exist for the treatment of the particle phase(s). We detail on our recent advances in the modeling of reactive mixing for organic nanoparticle precipitation. The Large Eddy Simulation approach is used for this purpose facilitated with an appropriate presumed-PDF reactive mixing model [1]. A study of inert and reactive mixing (see Figure 1) in a micro reactor as well as an investigation of different nucleation kernels is provided. This is done for the first time for a precipitation process involving organic nanoparticles. Also, we directly simulated particle transport processes in pharmaceutical multiphase reactors. This has been done using the Euler-Lagrange approach based on our previous work [2]. Specifically, we used a hard-sphere collision model and an appropriate stochastic model to account for particle-particle collisions and unresolved velocity fluctuations, respectively. Having this detailed information at hand, coalescence processes were studied directly. Also, collision statistics (see Figure 2) between two heterogeneous particle populations are reported for the first time using such a sophisticated particle tracking approach. References [1] S. Radl, D. Suzzi, J.G. Khinast: Fast Reactions in Bubbly Flows: Film Model and Micro-Mixing Effects, submitted to Industrial & Engineering Chemistry Research, 2010. [2] S. Radl, R. Sungkorn, D. Suzzi, J. Derksen, J.G. Khinast: High Performance Simulation of Bubbly Flows: Pushing the Limit by Using Conventional CFD and LBGK, Multi-Scale Modelling Symposium, Melbourne (Australia), 2009.

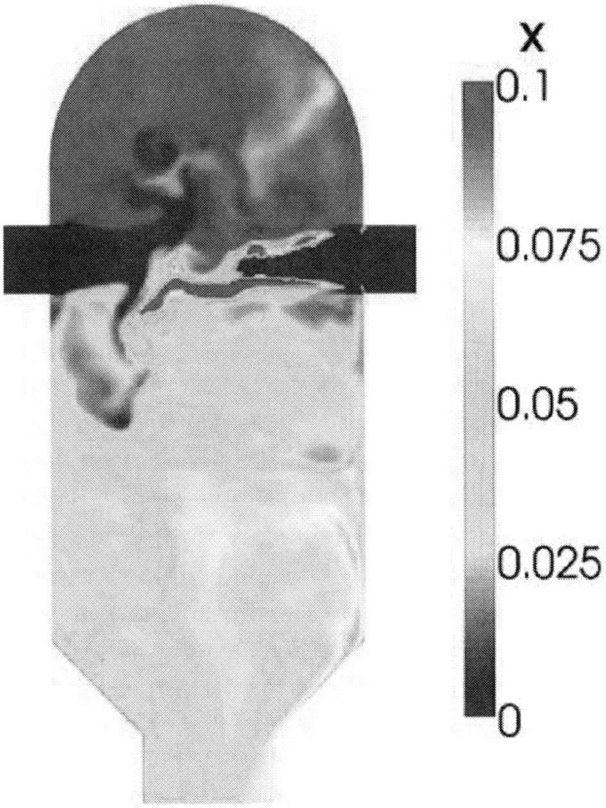

Figure 1: Distribution of the conversion in a microreactor for nanoparticle precipitation

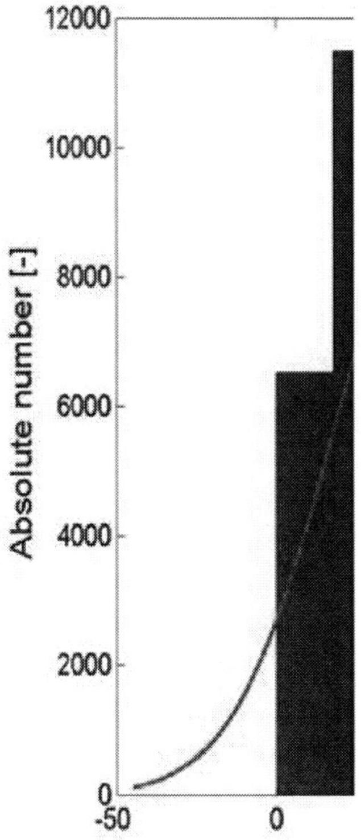

Figure 2: Directly bubble interactio

See more of this Session: Mixing in Multi-Phase Systems II

See more of this Group/Topical: North American Mixing Forum

(158f) Production of Pickering Emulsions in Industrial Equipment: a Stabilization Model

Emir Tsabet, Ecole Polytechnique de Montreal, Montreal, QC, Canada and **Louis Fradette**, URPEI, Chemical Engineering Department, Ecole Polytechnique de Montreal, Montreal, QC, Canada

Conventional emulsions are encountered in numerous industrial fields and since the pioneering work of Pickering and Ramsden a hundred years ago it has also been known that fine solid particles can be used as emulsifiers. It is also well established that correctly selected particles produce emulsions that are more stable than the ones stabilized with chemical surfactants. The characterization of these systems and their features have been investigated, such as the effect of the particles contact angle, their concentration, shape and size, the effect of aqueous phase pH and ionic strength and the effect of oil phase polarity and viscosity. In the light of the available work, it is obvious that it is possible to use particles to stabilize emulsions at the industrial scale. Unfortunately, the currently available information is not sufficient to design processes or to develop predictive models of the emulsions behavior. Being interested in emulsification process design, we are developing a methodology to predict the solid stabilized emulsions properties from the process conditions. The present work aims at developing a model describing the droplets stabilization on the basis of particle adsorption forces measurements using Atomic Force Microscopy. This approach allows us to identify the main parameters affecting the particles adsorption mechanism and therefore the stabilization rate, a condition to predict the droplets distribution size during the emulsification process.

See more of this Session: Mixing in Multi-Phase Systems II
See more of this Group/Topical: North American Mixing Forum

Transesterification of lipids with supercritical alcohols for biofuels production

Esteban A. Brignole
Planta Piloto de Ingeniería Química - Universidad Nacional del Sur - CONICET
8000 Bahía Blanca, Argentina

Since the early studies of Saka and Kusdiana on the noncatalytic transesterification of vegetable oils with supercritical alcohols, a large collection of reaction data for different vegetable oils and lipid-containing materials, with methanol or other alcohols, has been reported in the open literature. The wealth of information gathered in the last decade will be discussed to present general guidelines to carry out these processes under optimum conditions regarding alcohol/fatty oil molar ratio, operating temperature and pressure and contact time. Very promising have been the studies that indicate that neither water nor free fatty acids interfere with the transesterification process. On the contrary, it seems that their presence in the mixture fed to the reactor is beneficial for the transesterification reaction. This property and the absence of a catalyst simplifies the biodiesel purification, produces better quality glycerin, and much less waste than the conventional catalytic processes.

The experimental studies have progressed from closed batch reactors, to windowed reactors, to continuous reactors. The problems associated to reaction system density, phase conditions, thermodynamic modeling, and effect of cosolvents have all been given considerable attention. The conditions that cause degradation of the fatty esters have also been investigated. As an additional application the process has been used in the recovery and conversion to biodiesel of the lipids contained in algae cells.

Preliminary economic evaluations of large scale production processes have been made. These evaluations indicate that this technology has economic potential and is environmentally benign.

(232a) A Novel Slinger for Efficient Reflux Condensate Distribution in the Boiling Reactor

Kishore K. Kar[1], Madan Somasi[2], Richard F. Cope[1] and Hua Bai[3], (1)The Dow Chemical Co., Midland, MI, (2)The Dow Chemical Company, Midland, MI, (3)The Dow Chemical Company, Freeport, TX

In a multiphase continuous boiling reactor, often condensate from the overhead reflux condenser is fed back to the reactor through either a dedicated condensate return or the feed line. Despite some advantages to adding all the condensate into the feed line, a significant portion is purposely returned directly to the reactor, where a slinger distributes it over the free liquid surface. Commercial slingers consist of multiple straight, radial vanes atop a horizontal disc that rotates with the mixing shaft. Ideally, slinger rotation in the reactor headspace spreads liquid over the entire annular gap between the disc and the reactor wall. This creates a full liquid curtain, or "umbrella," that removes condenser-plugging particles from the upward-migrating vapor and rind-forming particles from the vessel wall. Unfortunately, the liquid from a standard slinger typically covers only a portion of the reactor cross-section. Therefore, it cannot prevent such plugging and rind formation. A novel slinger (KS-8) comprising eight curved radial vanes was developed and implemented in a Dow plant to overcome this liquid distribution limitation, and thus, improved asset reliability and utilization. This presentation addresses the key phases of the KS-8 development program: Ideation, Experimental, CFD Validation, and Implementation in the plant.

See more of this Session: Mixing in Process Industries
See more of this Group/Topical: North American Mixing Forum

AIChE 193457
2010, AIChE Annual Conference, North American Mixing Forum

Cooling Coil Effects On Blending in a Pilot Scale Tank

Robert A. Leishear, Savannah River National Laboratory, Aiken, SC

Mark D. Folwey, Savannah River National Laboratory, Aiken, SC

Michael R. Poirier, Savannah River National Laboratory, Aiken, SC

Timothy J. Steeper, Savannah River National Laboratory, Aiken, SC

Abstract

Blending, or mixing, processes in 1.3 million gallon nuclear waste tanks are complicated by the fact that miles of serpentine, vertical, cooling coils are installed in the tanks. As a step toward investigating blending interference due to coils in this type of tank, a 1/10.85 scale tank and pump model were constructed for pilot scale testing. A series of tests were performed in this scaled tank by adding blue dye to visualize blending, and by adding acid or base tracers to solution to quantify the time required to effectively blend the tank contents. The acid and base tests were monitored with *pH* probes, which were located in the pilot scale tank to ensure that representative samples were obtained. Using the probes, the hydronium ion concentration $[H^+]$ was measured to ensure that a uniform concentration was obtained throughout the tank. As a result of pilot scale testing, a significantly improved understanding of mixing, or blending, in nuclear waste tanks has been achieved. Evaluation of test data showed that cooling coils in the waste tank model increased pilot scale blending times by 200 % in the recommended operating range, compared to previous theoretical estimates of a 10 – 50% increase. Below the planned operating range, pilot scale blending times were increased by as much as 700 % in a tank with coils installed. One pump, rather than two or more, was shown to effectively blend the tank contents, and dual pump nozzles installed parallel to the tank wall were shown to provide optimal blending. In short, experimental results varied significantly from expectations.

This document was prepared in conjunction with work accomplished under Contract No. DE-AC09-08SR22470 with the U.S. Department of Energy. This work was prepared under an agreement with and funded by the U.S. Government. Neither the U. S. Government or its employees, nor any of its contractors, subcontractors or their employees, makes any express or implied: 1. warranty or assumes any legal liability for the accuracy, completeness, or for the use or results of such use of any information, product, or process disclosed; or 2. representation that such use or results of such use would not infringe privately owned rights; or 3. endorsement or recommendation of any specifically identified commercial product, process, or service. Any views and opinions of authors expressed in this work do not necessarily state or reflect those of the United States Government, or its contractors, or subcontractors.

Introduction

Project Overview

The Savannah River National Laboratory (SRNL) in conjunction with Savannah River Remediation (SRR) has completed some scaled blending tests using a pilot scale model of a nuclear waste tank to determine the flow rates and nozzle diameters for a fixed position, non-rotating, dual nozzle pump. The model was based on an 85 foot diameter SRS (Savannah River Site) waste tank (Tank 50H, Figures 1 and 2) and included tank internals such as pumps, 2 inch diameter cooling coils, and a central roof support column. The purpose of the testing was for SRNL to recommend design parameters to SRR for a blending pump as well as pump orientation and Tank 50H blending time. Presently, mixing pump motors are mounted at the top of waste tanks, and a 35 foot long drive shaft passes through the tank top to the mixing pump located below in the waste liquid. Further research is in progress for SRR, but this paper focuses on the initial pilot scale test results (Leishear, Fowley, and Poirier [1]).

Figure 1: Typical Mixing or Slurry Pumps Installed on the Top of an Underground Tank at Savannah River Site

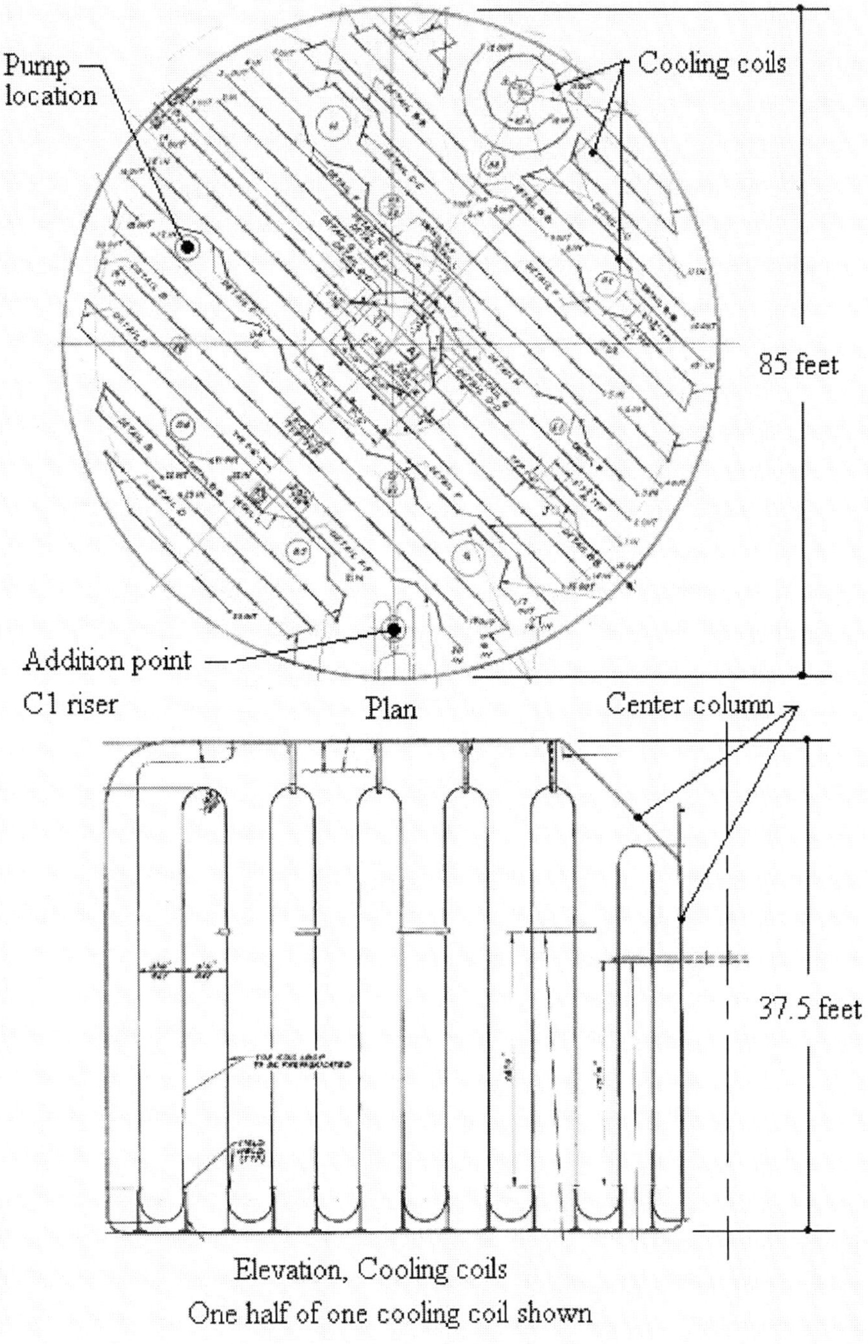

Figure 2: Tank 50H, Cooling Coil Arrangement

Pilot Scale Testing

To model blending processes, a series of tests were performed using small quantities of acid and base chemical additions, which were blended using three different pilot scale pump models. These three pump models provided different length and diameter characteristics to investigate the effects of these characteristics on blending. The pilot scale pump locations and chemical addition point locations were scaled down from Tank 50H. Pilot scale tests were used to evaluate blending effectiveness by blending nitric acid and sodium hydroxide solutions in a 40 inch deep by 94 inch inside diameter tank, which was scaled down from an 85 foot diameter nuclear waste tank (Figures 3 – 6). Dimensions were geometrically (linearly) scaled, and the pump discharge velocity remained constant at both scales to ensure that jet velocities were comparable at equivalent locations in both tanks. Concentrated solutions of acid and base (250 - 1000 milliliters) were added to 980 gallon batches of liquid (i.e., water and dilute sodium nitrate formed from acid-base reactions) at a location selected to match the addition point location in the waste tank (C1 riser). The tank contents were initially acidic or basic, and chemical additions changed the *pH* of the tank contents. That is, blending changed the contents from *pH* ≈ 4 to *pH* ≈ 10, or vice versa. The *pH* was monitored at several strategic locations and blending was considered complete when the *pH* readings converged to nearly constant values. Blending times were determined using these chemical changes.

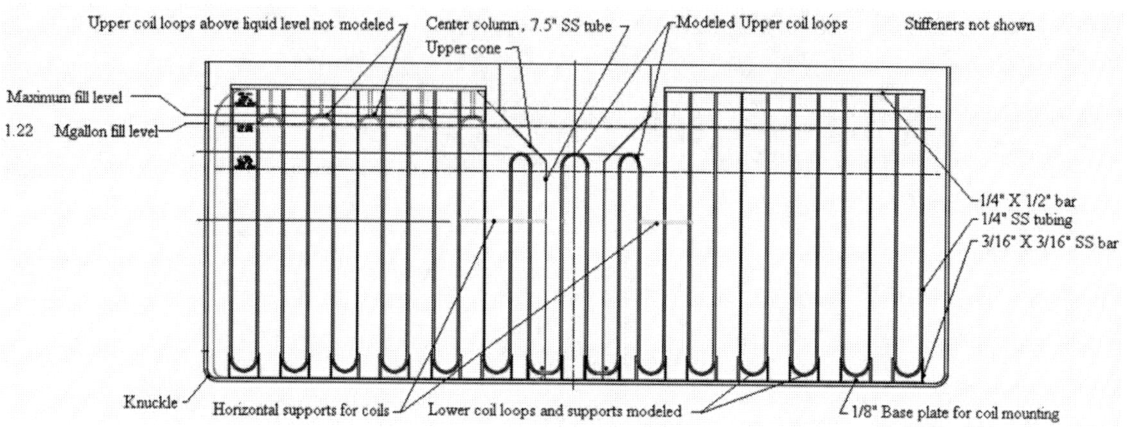

Figure 3: Pilot Scale Blending Tank, Elevation

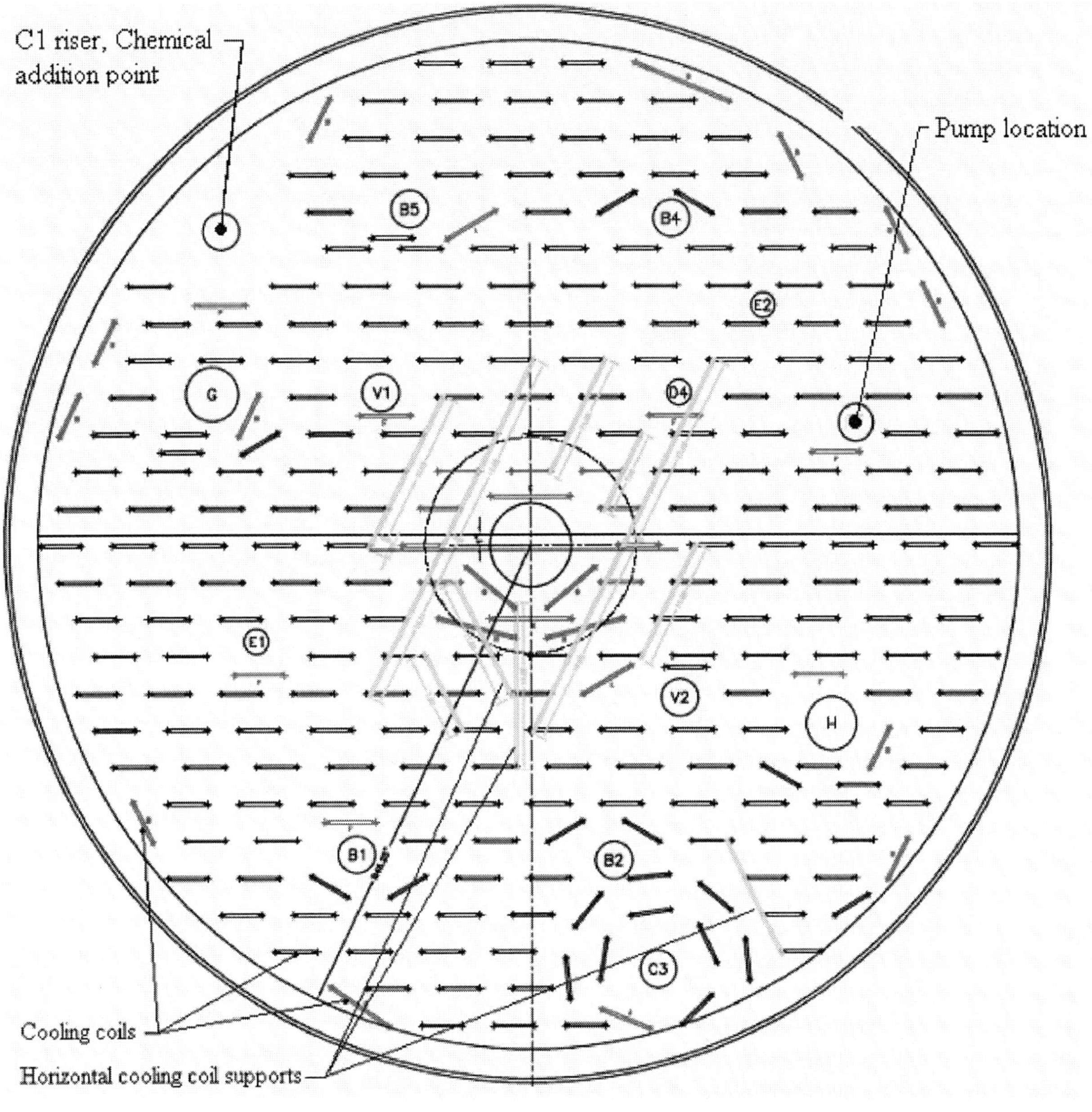

Figure 4: Cooling Coil Model, Cross Section

Figure 5: Modeled Cooling Coils and Center Column Assembly with Stiffeners and Lifting Lugs

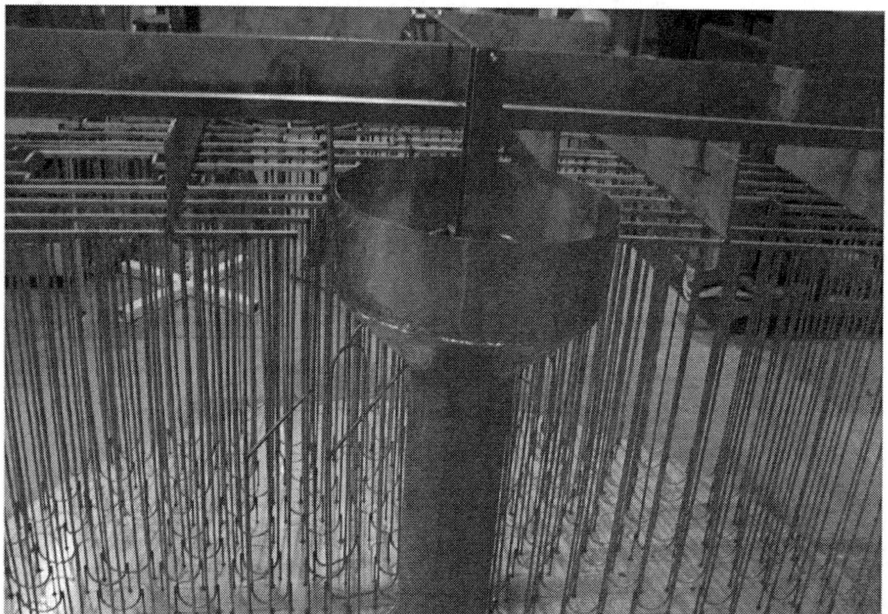

Figure 6: Cooling Coil Assembly and Center Column

Analysis

Blending Theory

The controlling factor or design parameter for blending is expressed as U_0D (feet^2 / second), where U_0 is the discharge velocity of a blending pump nozzle, and D is the inside diameter of the pump nozzle. Nozzle diameters and flow rates were used to vary U_0D. The diameters of the nozzles used in testing were scaled from 1-1/2", 2-1/4", and 3-5/8" full scale nozzles. Equations describing the blending time for miscible liquids in tanks mixed with single turbulent jets are typically of the general form

$$t = (C \cdot T^2) / (U_0 \cdot D) \qquad \text{Equation 1}$$

where t is the blending time, C is a constant, and T is the tank diameter.

This equation shows that the blend time is a function of U_0D, where C is typically in the range: $3.0 < C < 4.5$. The most recently predicted value for C was 3.0 (Grenville and Tilton [Refs. 2 and 3]). Grenville noted that this value for C was valid for tank volumes up to 3 million gallons. Experimentally, the constant was required for the pilot scale models, since this value is based on experimental data presented in the literature for tanks blended without a center column or coils, and a single jet nozzle with a centerline coincident to the tank radius. Accordingly, the 3.0 value for C provided only an estimate to find flow rates and pilot scale blending times. The value for this constant was investigated as this study extended this simplified blending equation to a tank with dual nozzles in a tank with or without cooling coils. In short, Equation 1 was assumed to be valid for blending in pilot scale testing whether or not cooling coils and a center column were installed, even though previous testing had not been performed using cooling coils. This study showed that this simple equation was consistent with test results, even though the value of C varied for tanks with and without coils, and C is also affected by the number and location of nozzles.

Blending Performance and Blending Times

To quantify blending performance, blending times were determined using commercial 95% blending criteria. The Hydronium ion concentrations $[H^+]$ were calculated from pH measurements and normalized to establish mixing times for 95% mixing (Paul, et.al [Ref. 4]). The 95% mixing criteria is a generally accepted criterion which defines the time following the addition of a tracer at which the concentrations throughout the tank are within ± 5 % of the bulk concentration. Normalization is a common practice for empirically quantifying mixing using concentration measurements. The 95% mixing time provided blending acceptance criteria, but lacked accuracy to quantify chemical concentrations throughout blended liquids.

From Paul, et. al. [Ref. 4], *pH* probes are commonly used to establish 95% blending times, which are determined from concentrations after adding a reactive tracer. To do so, a normalized concentration is calculated, where

$$C_i' = \frac{C_i - C_0}{C_\infty - C_0} \qquad \text{Equation 2}$$

where C'_i equals the normalized concentration, C_i equals the measured variable concentration, C_0 equals the initial concentration, and C_∞ equals the final equilibrium concentration. The 95% blending time equals the time required for the normalized probe output to reach and remain within 95 to 105%. Equations 3 to 8 provide relationships between concentration and *pH*. For the *pH* probe response, Equation 2 is rewritten as

$$\text{Normalized} \quad \left[H^+\right] = \frac{\left[H^+\right]_i - \left[H^+\right]_0}{\left[H^+\right]_\infty - \left[H^+\right]_0} \qquad \text{Equation 3}$$

and

$$pH = -\log\left[H^+\right] \qquad \text{Equation 4}$$

$$\left[H^+\right] = 10^{-pH} \qquad \text{Equation 5}$$

$$pOH = -\log\left[OH^-\right] \qquad \text{Equation 6}$$

$$\left[OH^-\right] = 10^{-pOH} = 10^{pH-14} \qquad \text{Equation 7}$$

$$pH + pOH = 14 \qquad \text{Equation 8}$$

Typical acceptance criteria for good blending in process industries are defined by 95% blending (normalized $[H^+]$), where a typical process is paint mixing. For pharmaceutical industries, where blending is more critical, 99% blending is sometimes used, where the normalized $[H^+] = 0.99 \rightarrow 1.01$. Appendix C shows that 99% blending is not recommended with the commercially used instrumentation, even though high quality instruments were used for this research (Hach, Inc.) to measure 95% blending as the recommended acceptance criteria for this research. Additional discussion of probe uncertainty and solution buffering effects follow, since they are related to the determination of 95% blending.

In a few cases, tests were prematurely stopped before reaching the 95% mixing time. In those cases, the blending time was estimated using

$$t = t_{95} = t_n \cdot \frac{\ln\left(1 - \frac{95}{100}\right)}{\ln\left(1 - \frac{n}{100}\right)} \qquad \text{Equation 9}$$

where t_{95} is the 95 % blending time, n is the percent blended, and t_n is a measured experimental blending time (Paul, et.al. [Ref.4]). This approximation assumes that the normalized concentration converges to the value of 1 in a smooth logarithmic fashion. In short, some tests were terminated prematurely, and Equation 9 was used to predict the estimated blending time if the test had been run to completion. Blending times obtained from this equation are approximate only, but provide insight into blending. A more detailed test equipment description is prerequisite to a discussion of test results.

Pilot Scale Equipment Description

Test Setup

The overall test setup is shown in Figure 7. The flow rate through the blending pump, Pump 1, was controlled by a variable frequency drive (VFD, Figure 8). The VFD provided the volumetric flow for blending through the pilot scale nozzles in the tank. . F1 and F2 are the flow meters used to measure and control flow rates through the nozzles, and probes 1 – 6 are *pH* probes. Data from the probes, the thermocouple (T1), the pump, and the flow meters were recorded by the data acquisition system (DAS, Figure 9).

Valves, V6 and V7, were used to add dye, acids, or bases to the tank for blending in a controlled, repeatable method. The miscible fluid to be blended was added to a funneled container, or hopper, above these valves, one of the valves was opened, and the acid or base was introduced below the surface of the tank contents by a ¼ inch diameter tube between the hoppers and tank contents. To eliminate transient blending concerns with respect to modeling and minimize testing, acids and bases were added to the tank after fluid flow in the tank reached steady state conditions. That is, all experimental blending times were provided from a steady state mixing condition. Proportional integral control (PI) was used to ramp up to steady state and maintain a constant flow rate through the pumps for pilot scale testing.

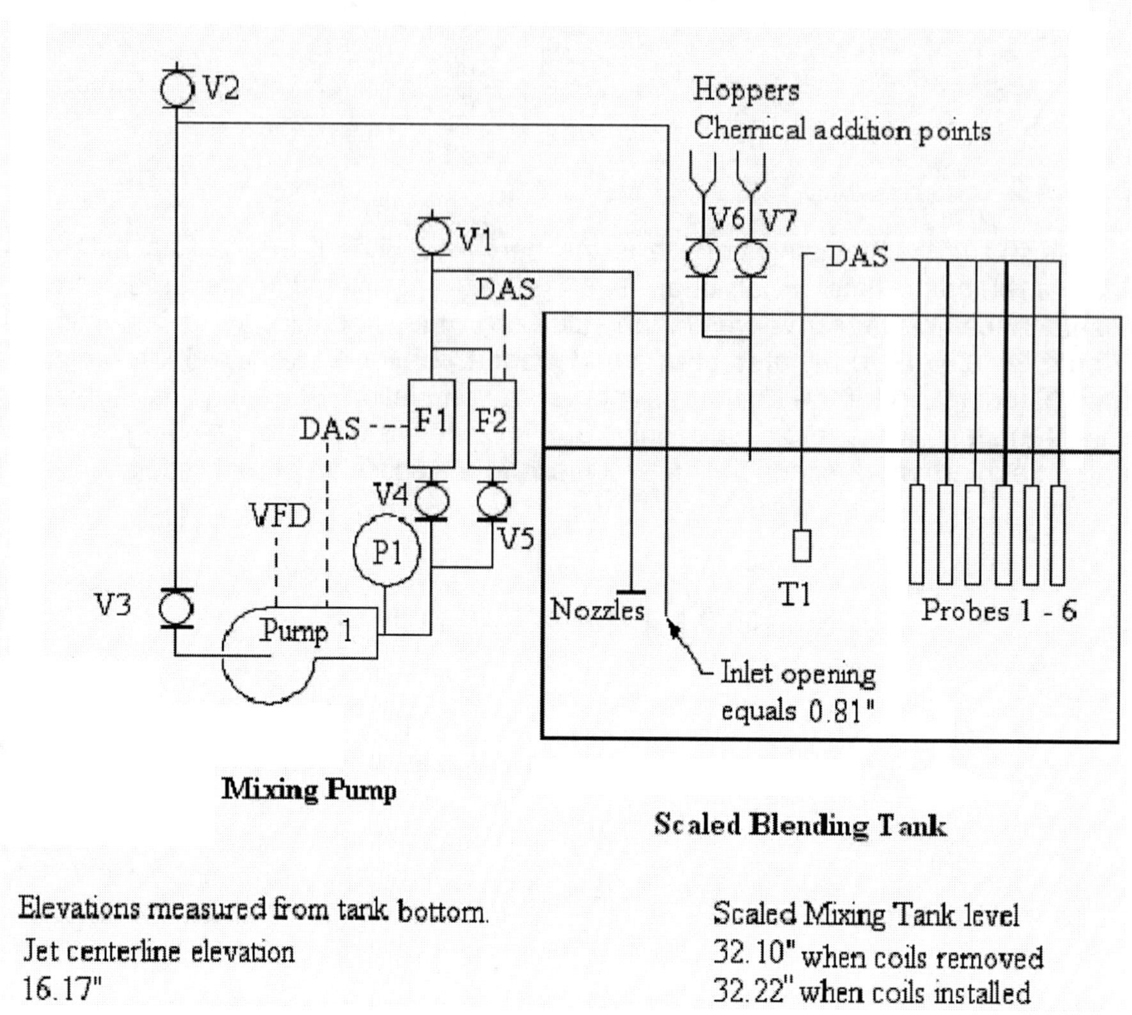

Figure 7: Test Schematic

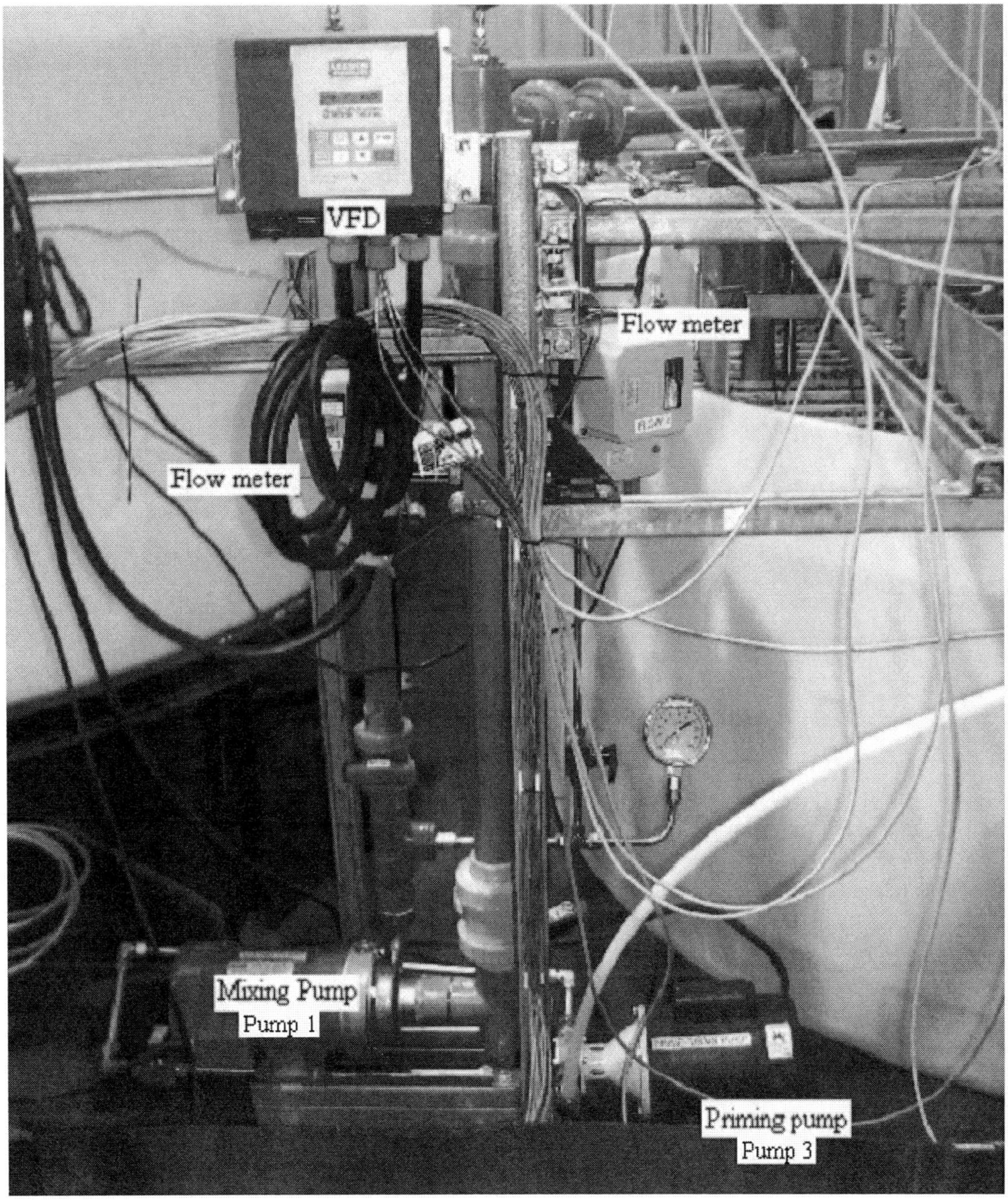

Figure 8: Mixing Pump for Pilot Scale Blending Tank

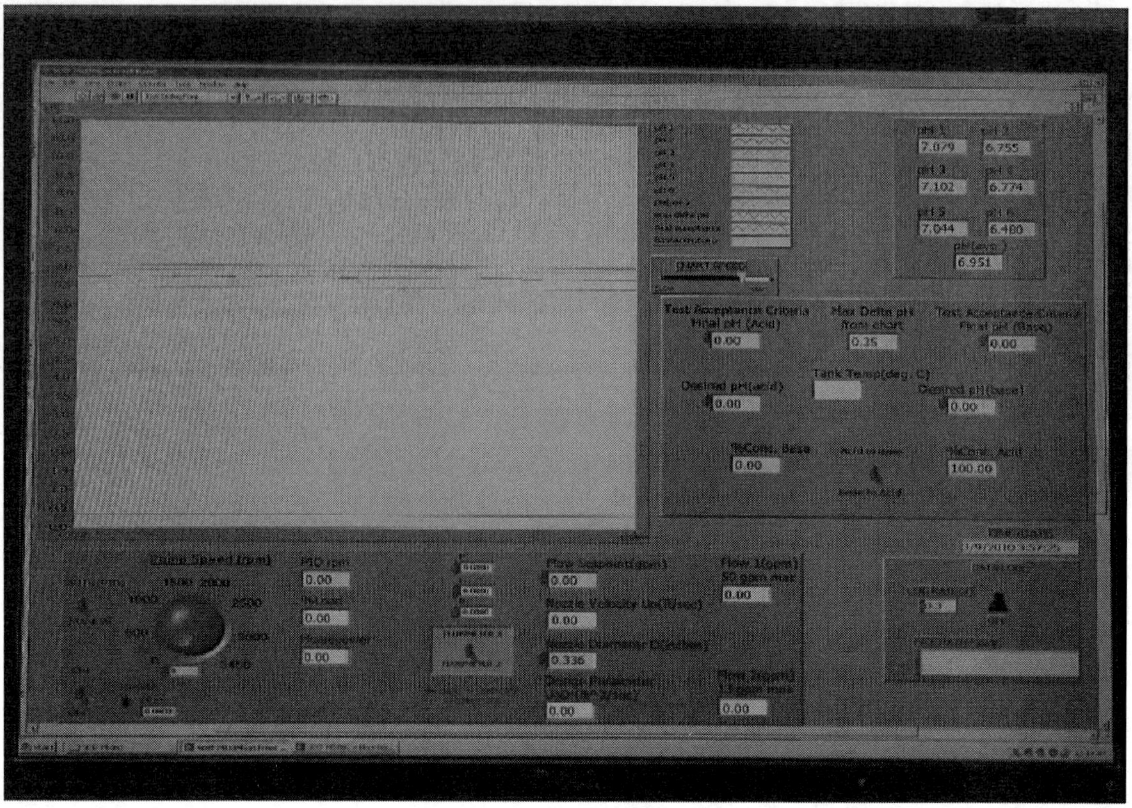

Figure 9: DAS Display

Pilot Scale Modeling

For some testing, simulated cooling coils were installed, but for other testing, coils were removed to quantify the effects of cooling coils on blending (Figures 10 and 11). The center column in the pilot scale tank was installed during all tests.

Before acid and base testing, a blue dye was added to the pilot scale test tank to qualitatively visualize blending and provide an indication of the slowest and fastest blended areas to determine optimal locations for *pH* probes (Figure 12). For acid and base testing, pump parameters were varied to investigate blending effects.

Parameters considered were nozzle length, nozzle diameter, nozzle velocity, and pump orientation. Pump orientations included nozzle positions parallel to the tank wall (referred to as the 0° position), perpendicular to the tank wall (90°), and at an angle of 45° to the tank wall (Figure 12). Pump nozzle designs are referred to as Standard, long Quad, and short Quad, which are facility specific references not explained herein. All tests were performed with the pump nozzles located at the mid-height of the test fluid in the pilot scale tank to investigate pump nozzle performance, and the pump suction was modeled by a drilled hole in the bottom of a pipe cap (Figures 13 and 14).

Once the test equipment was assembled, blending times were determined. In addition to calculating blending times, various effects on blending tests were considered, such as chemical buffering within the test fluids and diffusion.

Figure 10: Assembled Tank, With Cooling Coil Models

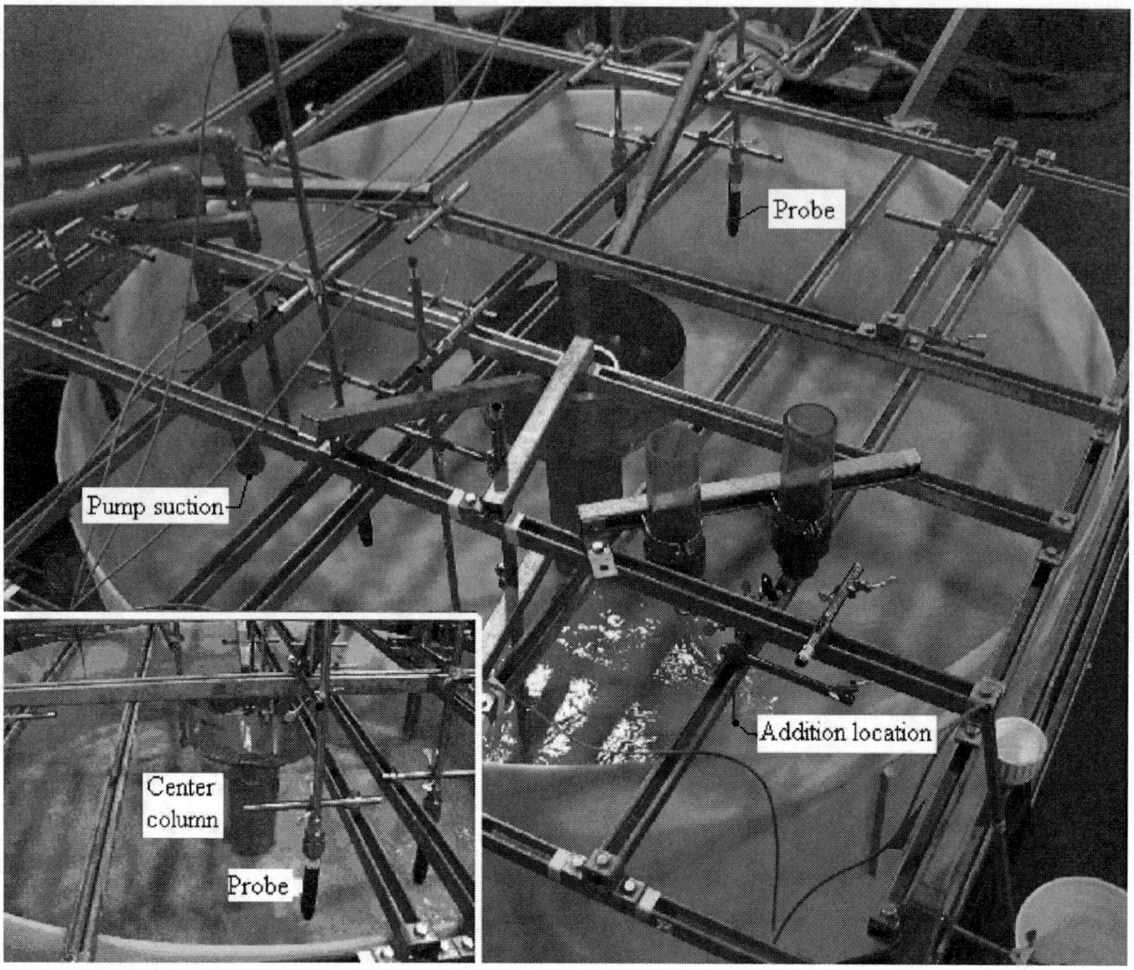

Figure 11: Assembled Tank, With Cooling Coil Models Removed

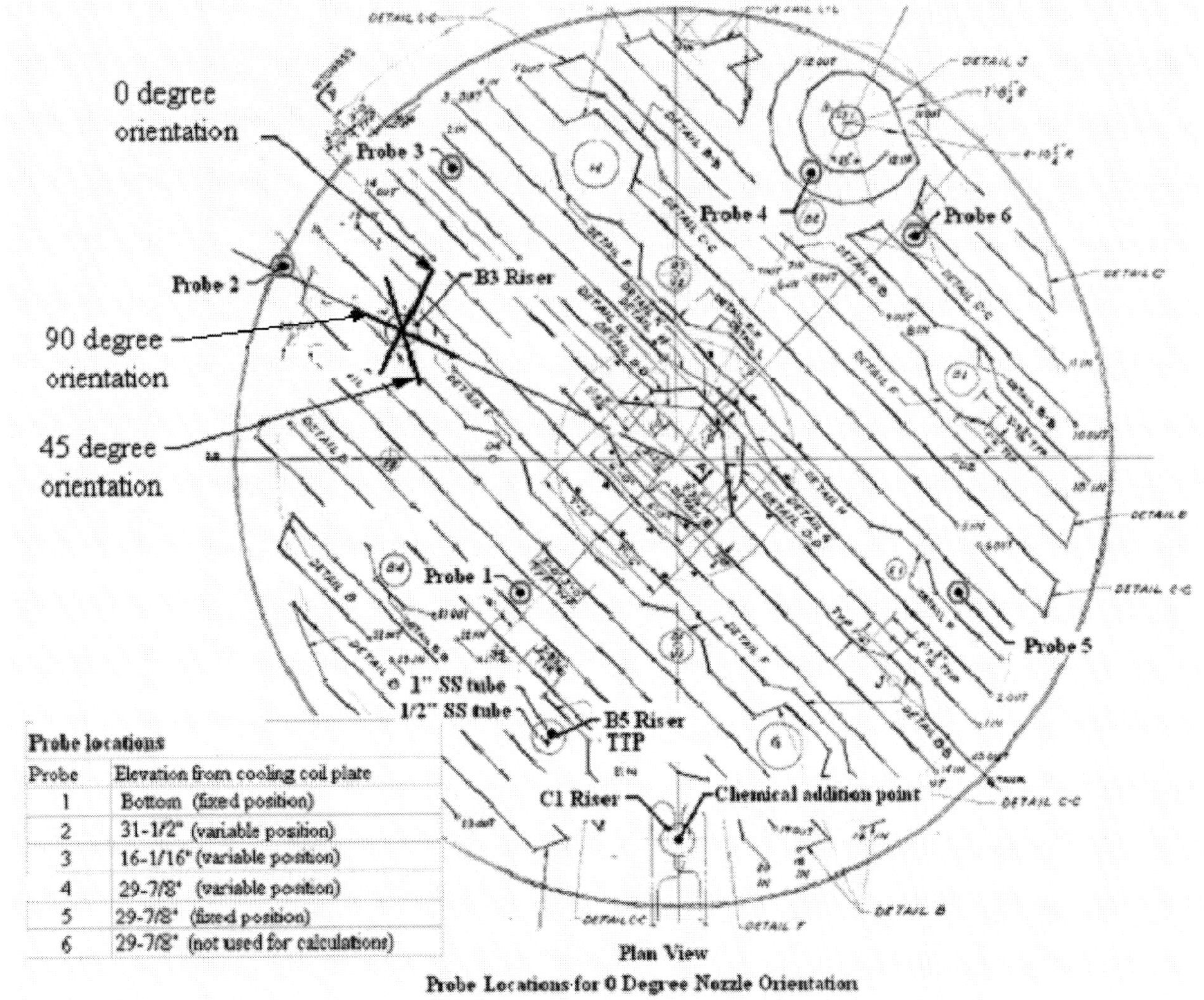

Figure 12: Nozzle Orientation and pH Probe Locations

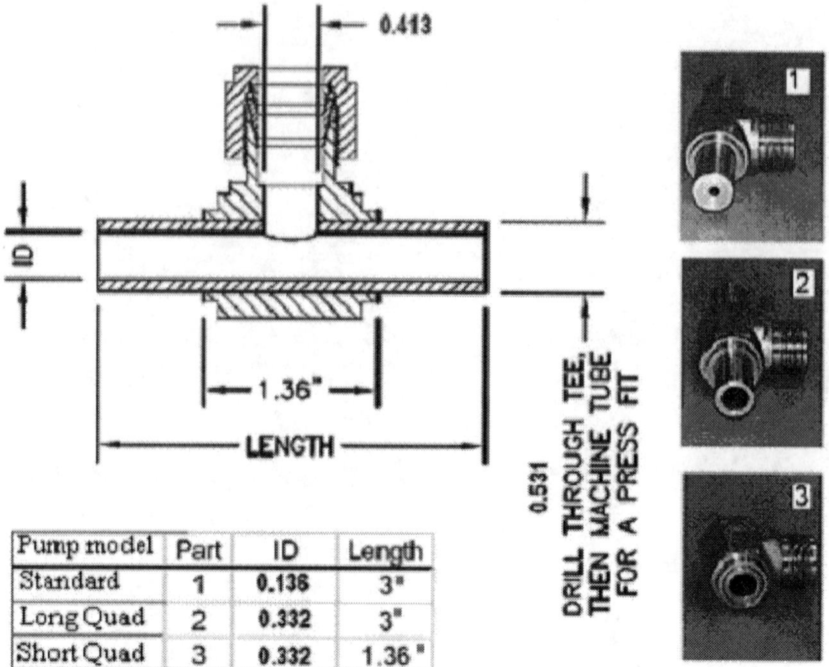

Pump model	Part	ID	Length
Standard	1	0.136	3"
Long Quad	2	0.332	3"
Short Quad	3	0.332	1.36"
SDI model	4	0.209	3"

Figure 13: Pump Models for Discharge Nozzles / Jets

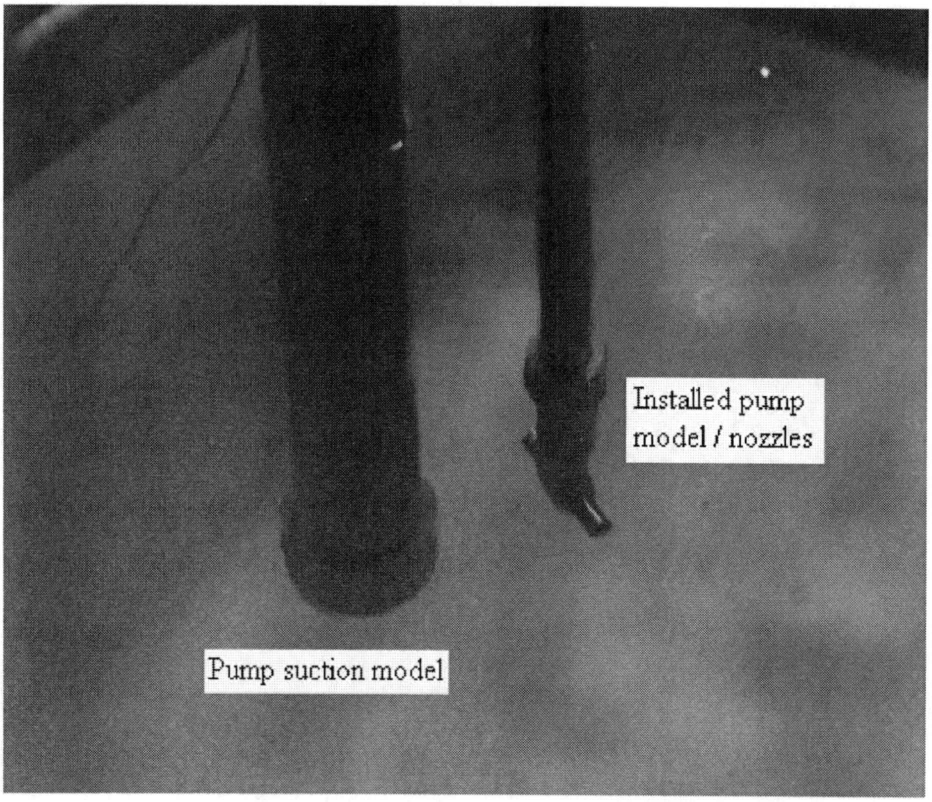

Figure 14: Pump Suction Inlet and Pump Nozzles

Results

Summary of Results

All test results are summarized in Table 1, where each of the parameters varied during testing is listed. A convenient testing term is introduced in the table, which is premature remixing. Between tests, the pump speed was increased to remix the tank contents. In those tests that were stopped before 95 % blending was achieved, the tank was said to be prematurely remixed. All tests were performed at controlled room temperatures, which varied between 18 and 21° C. The first four tests used 250 milliliters of blue dye added to the hopper at the C1 riser location, and the remaining tests used 250 and 500 milliliters of either acid (nitric acid) or base (sodium hydroxide) for additions to the tank hoppers. The hoppers are shown at the addition points in Figures 10 and 11. Figures 15 and 16 provide the results from Table 1 in graphic formats. A summary of the test results determined from Table 1 is followed here by some discussion of supporting tests. Again, further detail is available (Leishear, Fowley, and Poirier [1]).

To summarize test results:

- Pilot scale blending times were significantly affected by cooling coil installation. Blending times in a tank with coils were twice the blending times for a tank without coils, within the recommended range of operation. Below the recommended range of operation the basic fluid mechanics of blending is not understood, and blending times for a tank with coils was as much as seven times the blending time for a tank without coils.
- A minimum pilot scale U_0D = 0.47 ft^2 /second is recommended for effective blending (Equation 1). Below this recommended lower limit, flow characteristics change significantly and the scaling relationships used in this work are inapplicable (Figure 16).
- Data affected by premature remixing is below the recommended design condition of U_0D = 0.47 ft^2 /second.
- Dye tests typically had apparent blending times less than a third of the blending times quantitatively found during *pH* testing for comparable conditions.
- Diffusion was very slow when compared to blending times, and consequently had a negligible effect on blending.
- Pilot scale blending times varied by more than 700% for apparently comparable conditions.
- For *pH* tests, pilot scale blending times were independent of initial and final concentrations of acid or base, the concentration range during an acid or base test, or the initial acidic or basic condition in the tank when testing started.
- The effects of nozzle lengths on blending were inadequately quantified. Only two nozzle lengths were investigated for the Quad Volute pump, and only one test was performed on the alternative nozzle length.

- The 0° nozzle position is recommended.
- Perhaps due to limited data, larger diameter Quad nozzle pilot scale blending times are slightly shorter than smaller diameter Standard nozzle blending times for a tank with coils and comparable U_0D.
- Nozzle diameter effects were not investigated outside the range of the Standard and Quad nozzle diameters (1-1/2" – 3-5/8" scaled down to 0.138 and 0.334" respectively). At smaller diameters, conclusions with respect to U_0D and blending times may be questionable.
- Blending times were significantly shortened when a dye was added to the pump suction, instead of at the addition point hopper. The addition point location in the tank may affect blending times.
- A 95% blending time criteria was validated for use in test results, but a 99% blending time could not be obtained with available equipment.

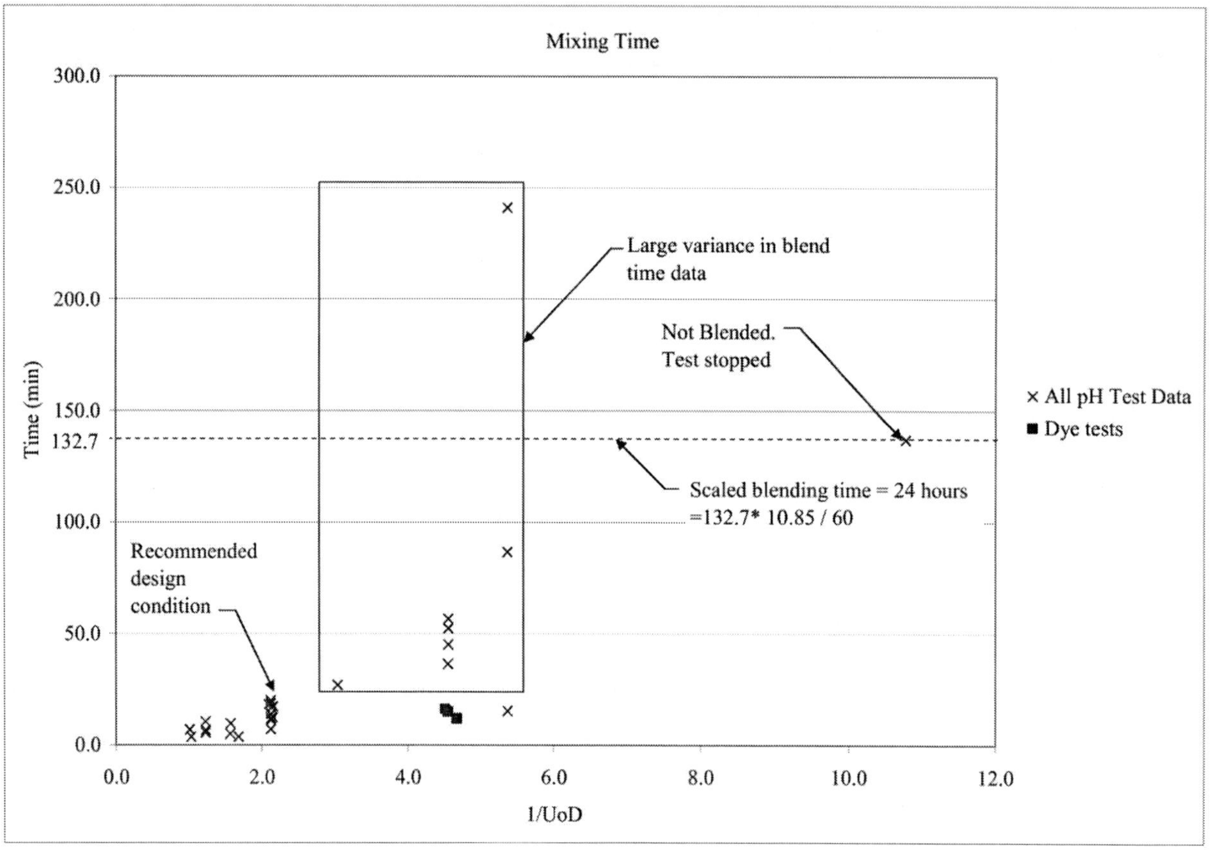

Figure 15: Measured Blending Times for All Tests

Table 1: Test Results

Test	Test type, Nozzle design	Fluid additions / strikes	Coils	Blue Dye	Nozzle position (deg)	D (in)	Uo (ft/s)	UoD (ft2/s)	Q (gpm)	Nozzle Re	Start cond	Start pH	End pH	Theoretical Time, no coils (min) t(theoretical)	Blending Time (min) t(measured)	Time ratio	Pilot scale UoD, relative standard deviation, %	Notes
1	Blending, dye, Quad	Dye	Y	Y	0	0.334	7.90	0.22	4.31	20,428	DI water	N/A	N/A	17.3	15	0.87	0.023	
2	Blending, dye, Quad	Dye	Y	Y	90	0.334	7.97	0.22	4.35	20,596	DI water	N/A	N/A	17.2	16.3	0.95	0.025	
3	Blending, dye, Quad	Dye	Y	Y	45	0.334	7.70	0.21	4.20	19,908	DI water	N/A	N/A	17.7	12	0.68	0.023	
4	Diffusion, dye	Dye	Y	Y	N/A	N/A	0	0	0	0	DI water	N/A	N/A	N/A	3.5 hours	N/A	N/A	
5	Diffusion, base	Base	Y	N	N/A	N/A	0	0	0	0	Acid	9.96	N/A	N/A	> 13.3 hours	N/A	N/A	
6	Diffusion, acid	Acid	N	N	N/A	N/A	0	0	0	0	Base	9.92	N/A	N/A	> 38.5 hours	N/A	N/A	
7	Blending, Quad	Base	Y	N	0	0.334	7.89	0.22	4.31	20,411	Acid	5.5	8.6	17.3	52.3	3.02	0.007	Premature remix
8	Blending, Quad	Acid	Y	N	0	0.334	7.89	0.22	4.31	20,402	Base	9.97	9.66	17.3	45.0	2.60	0.003	Premature remix
9	Blending, Quad	Acid	Y	N	0	0.334	7.90	0.22	4.31	20,428	Base	9.66	9.05	17.3	36.4	2.10	0.003	Premature remix
10	Blending, Quad	Base	Y	N	0	0.334	16.93	0.47	9.24	43,765	Acid	5.94	10	8.1	28.7	3.55	0.007	Premature remix
11	Blending, Quad	Acid	Y	N	0	0.334	16.94	0.47	9.24	43,791	Base	9.98	3.93	8.1	18.6	2.30	0.004	
12	Blending, Quad	Acid	Y	N	0	0.334	16.94	0.47	9.25	43,803	Base	10.22	8.7	8.1	11.0	1.36	0.005	
13	Blending, Quad	Base	Y	N	0	0.334	16.94	0.47	9.24	43,791	Acid	3.92	9.78	8.1	11.9	1.47	0.004	
14	Blending, Quad	Base	Y	N	0	0.334	29.29	0.82	15.99	75,738	Acid	4.27	10.21	4.7	6.6	1.41	0.054	
15	Blending, Quad	Base	Y	N	0	0.334	29.38	0.82	16.03	75,958	Acid	5.81	9.68	4.7	10.4	2.24	0.058	
16	Blending, Quad	Acid	Y	N	0	0.334	29.38	0.82	16.03	75,958	Base	9.97	4.23	4.7	5.5	1.18	0.073	
17	Blending, Short Quad	Base	Y	N	0	0.33	17.35	0.48	9.25	44,326	Acid	3.96	9.96	8.0	18.1	2.27	0.004	
18	Blending, Standard	Base	Y	N	0	0.134	20.93	0.23	1.84	21,713	Acid	4.06	10.07	16.3	116.9	7.19	0.007	Premature remix
19	Blending, Standard	Acid	Y	N	0	0.134	41.64	0.46	3.66	43,193	Base	9.96	4.05	8.2	17.3	2.11	0.009	
20	Blending, Quad	Acid, dye	Y	Y	90	0.334	16.94	0.47	9.24	43,791	Base	9.91	3.94	8.1	20.0	2.48	0.004	
21	Blending, Quad	Base	Y	N	90	0.334	16.94	0.47	9.24	43,791	Acid	3.95	9.92	8.1	13.7	1.70	0.004	
22	Blending, Quad	Acid	Y	N	45	0.334	3.34	0.09	1.82	8,624	Base	10.12	4.03	41.0	136.80	none	0.011	Not blended
23	Blending, Quad	Acid	Y	N	45	0.334	6.70	0.19	3.66	17,325	Base	9.87	4.14	20.4	149.9	7.35	0.003	Blending incomplete
24	Blending, Quad	Base	Y	N	45	0.334	7.89	0.22	4.31	20,402	Acid	3.89	9.83	17.3	56.5	3.26	0.003	Premature remix
25	Blending, Quad	Acid	Y	N	45	0.334	11.90	0.33	6.50	30,771	Base	9.98	3.86	11.5	31.7	2.76	0.003	
26	Blending, Quad	Acid	Y	N	45	0.334	16.94	0.47	9.24	43,791	Base	10.11	3.96	8.1	16.8	2.08	0.004	
27	Blending, Quad	Base	Y	N	45	0.334	22.89	0.64	12.49	59,176	Acid	4.06	9.95	6.0	9.5	1.59	0.024	
28	Blending, Quad	Base	Y	N	45	0.334	35.71	0.99	19.49	92,339	DI water	6.45	9.95	3.8	6.8	1.78	0.065	
29	Blending, Quad	Acid	N	N	45	0.334	6.70	0.19	3.66	17,325	Base	9.97	5.63	20.4	15.2	0.74	0.003	
30	Blending, Quad	Base	N	N	45	0.334	6.70	0.19	3.66	17,325	Acid	4.02	10.13	20.4	23.2	1.14	0.003	Blending incomplete
31	Blending, Quad	Acid	N	N	45	0.334	11.85	0.33	6.47	30,629	Base	10	4.05	11.5	12.3	1.07	0.004	
32	Blending, Quad	Base	N	N	45	0.334	16.74	0.47	9.14	43,273	Acid	4.1	9.97	8.2	4.9	0.60	0.021	
33	Blending, Quad	Base	N	N	45	0.334	22.98	0.64	12.54	59,422	Acid	4.02	9.99	5.9	3.6	0.61	0.022	
34	Blending, Quad	Acid	N	N	45	0.334	34.99	0.97	19.10	90,477	Base	9.93	4	3.9	4.0	1.03	0.029	
35	Blending, Standard	Acid	N	N	45	0.134	9.22	0.105	0.81	9,560	Base	9.96	3.99	37.0	86.6	2.34	0.006	Premature remix
36	Blending, Standard	Base	N	N	45	0.134	20.93	0.23	1.84	21,713	Acid	3.98	9.95	16.3	15.6	0.96	0.007	
37	Blending, Standard	Acid	N	N	45	0.134	42.30	0.47	3.72	43,878	Base	10.11	3.96	8.1	7.0	0.87	0.051	
38	Blending, Standard	Base	N	N	45	0.134	52.84	0.60	4.64	54,817	Acid	3.9	9.91	6.4	3.6	0.56	0.052	
39	Blending, 0.209 " nozzles	Dye	Y	Y	0	0.209	46.51	0.81	9.94	75,251	DI water	N/A	N/A	4.6	0.42	0.09	N/A	Dye added to suction pipe

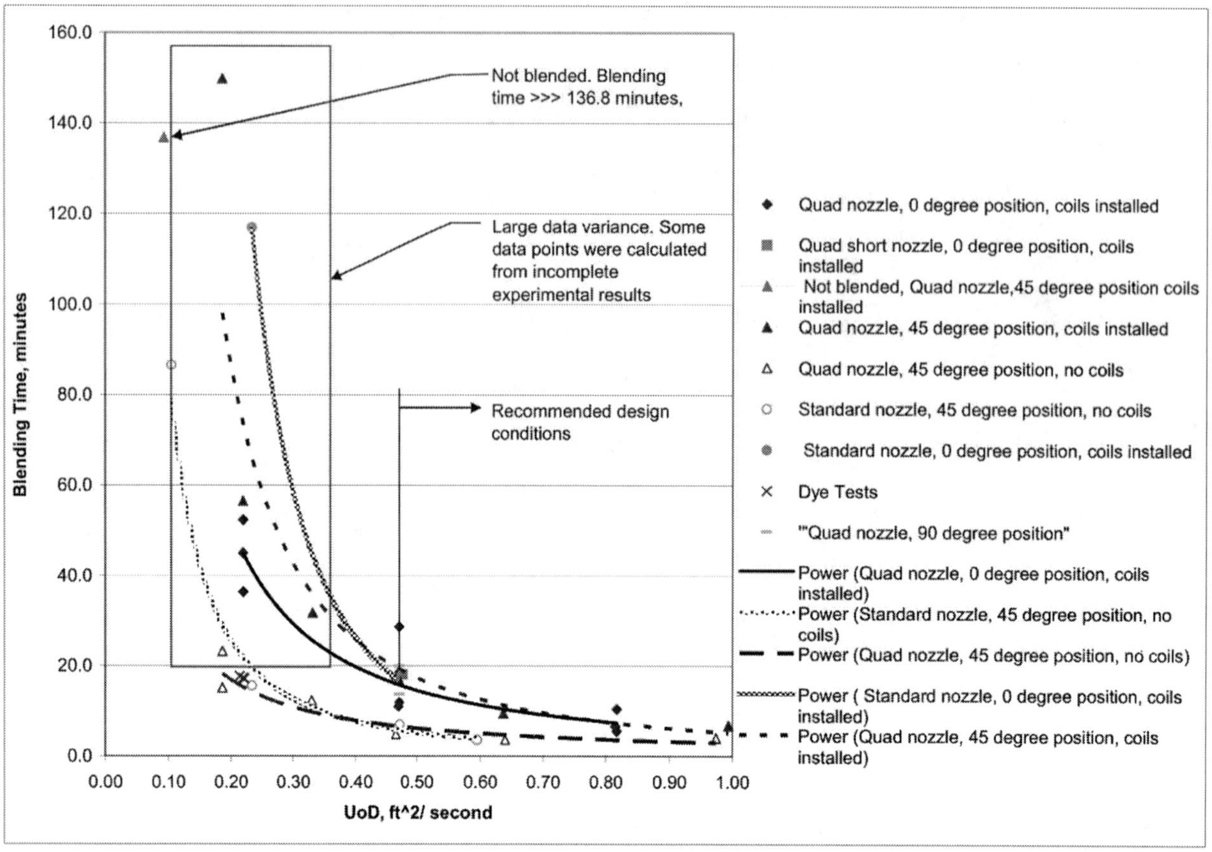

Figure 16: U_0D vs. Pilot Scale Blending Times for Acid and Base Tests

Dye Tests

The addition of dye to the hopper for one of the tests is shown in Figure 17, and the distribution of the dye across the tank floor is shown in Figure 18. Although the distribution varied somewhat for different nozzle orientations, this figure depicts the typical spread of dye in the pilot scale tank. For one test (Test 39), dye was added directly to the pump suction piping, and the tank blended in about 30 seconds, which was much faster than other tests with comparable U_0D. This result indicated that blending time may also be affected by the location of the chemical addition point.

During later *pH* testing, the acids were lower viscosity than bases and spread across the entire floor faster, while the thicker, higher viscosity bases tended to flow directly toward the pump suction along the tank floor. The acid had a higher viscosity and density than water, while the base had an even higher viscosity and density. In later acid-base testing, the *pH* probes indicated that at higher flows, the acids and bases actually spread across the air / water surface, rather than dropping immediately to the tank floor, when a *pH* probe at the surface responded to *pH* changes before the probe on the tank floor responded. Regardless of acid or base addition, the results showed that the selection of an acid or base addition had a negligible effect on pilot scale blending times.

Dye Addition Started, C1 Riser

Dye Addition Complete, C1 Riser

Dye Addition Near Completion, C1 Riser

Dye Spreading Across the Tank Floor, B3 Riser, Near the Pump

Figure 17: Dye Addition at C1 Riser (Test 1)

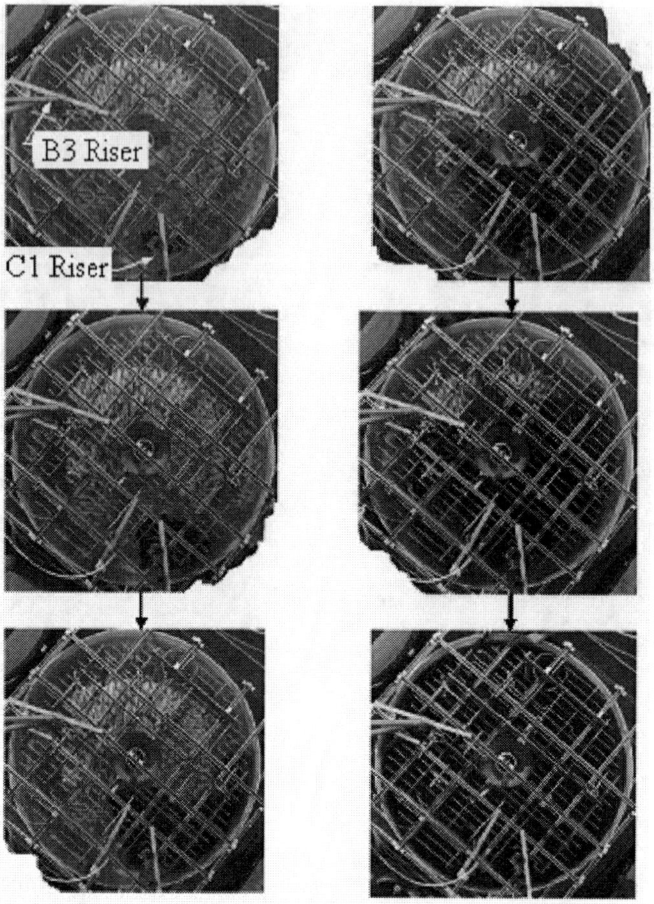

SDI, Tank 50 Model, Plan View, Blue Dye Blending, Scaled Quad Pump Using Scaled Nozzle Diameters, Nozzles Parallel to Tank Wall

Figure 18: Typical Dye Distribution in the Pilot Scale Tank (Test 1)

Acid and Base Testing Using pH Measurements

Test 11 is provided as an example of test data used to establish the blending times listed in Table 1. The raw data was displayed on the DAS in the format shown in Figure 19 during testing. The zero time, $t = 0$, equaled the time at which the valve on the hopper at the C1 riser location was opened to release acid into the tank. The pump was already operating at the selected speed and flow rate. Probe 1 was located near the upper surface of the tank fluid, while the other probes were located near the tank bottom, as shown in Figure 12.

Several steps were taken to calculate bending times. First, *pH* data was converted to hydronium ion concentrations as shown in Figures 20 and 21. Then, a bias, or offset gain, was applied to the data to eliminate the offset between probes, and ensure that the data converged, or collapsed, to a common datum. Finally, the data was normalized to establish pilot scale blending times using Equation 2, as shown in Figure 22. Most test results are similar, except the limiting case where the tank contents did not completely mix, which is shown in Figure 23. Numerous test details, and plots for all tests are available (Leishear, Fowley, and Poirier [1]), where blending times for all tests were similarly obtained. Details of diffusion, buffering, and instrument uncertainty also require some discussion.

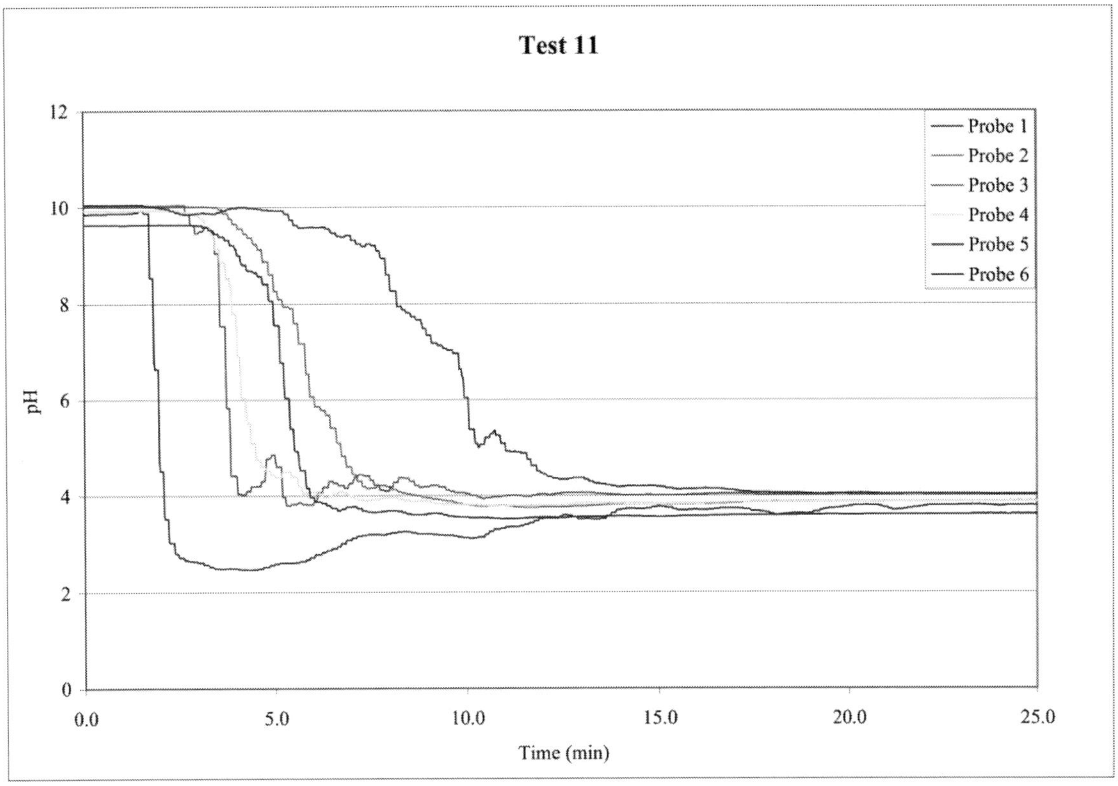

Figure 19: Raw pH Data for Analysis

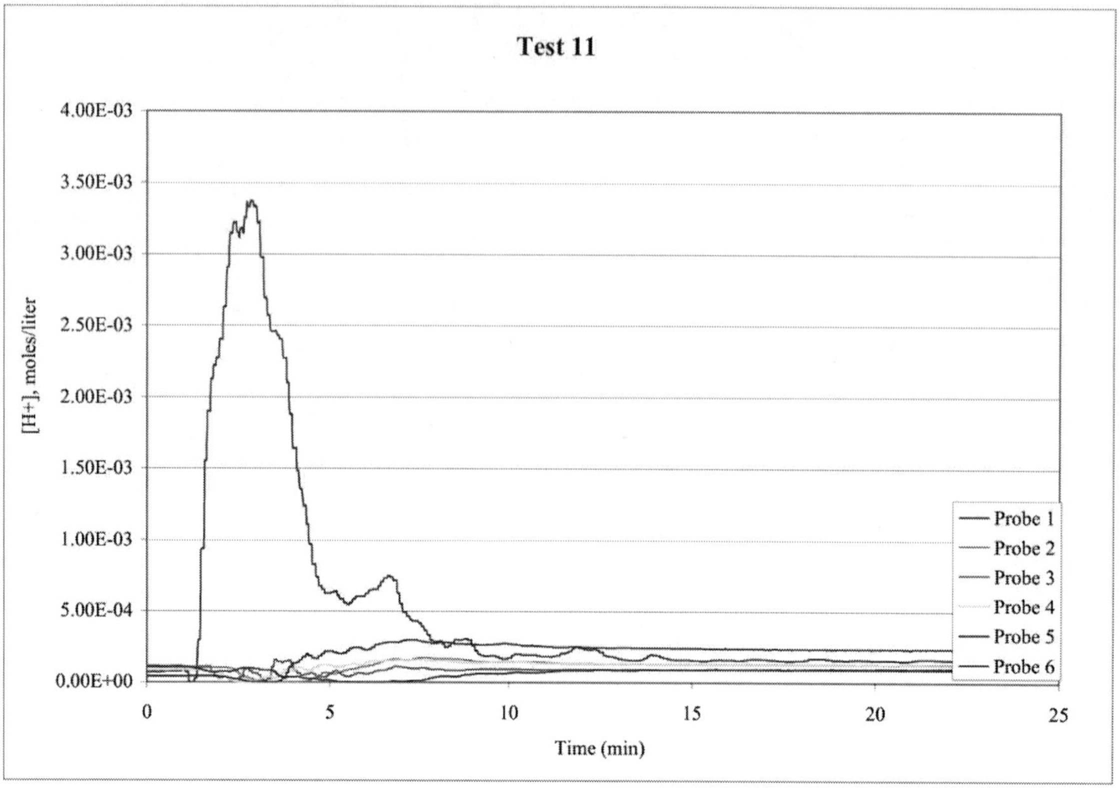

Figure 20: Concentration Data for Analysis

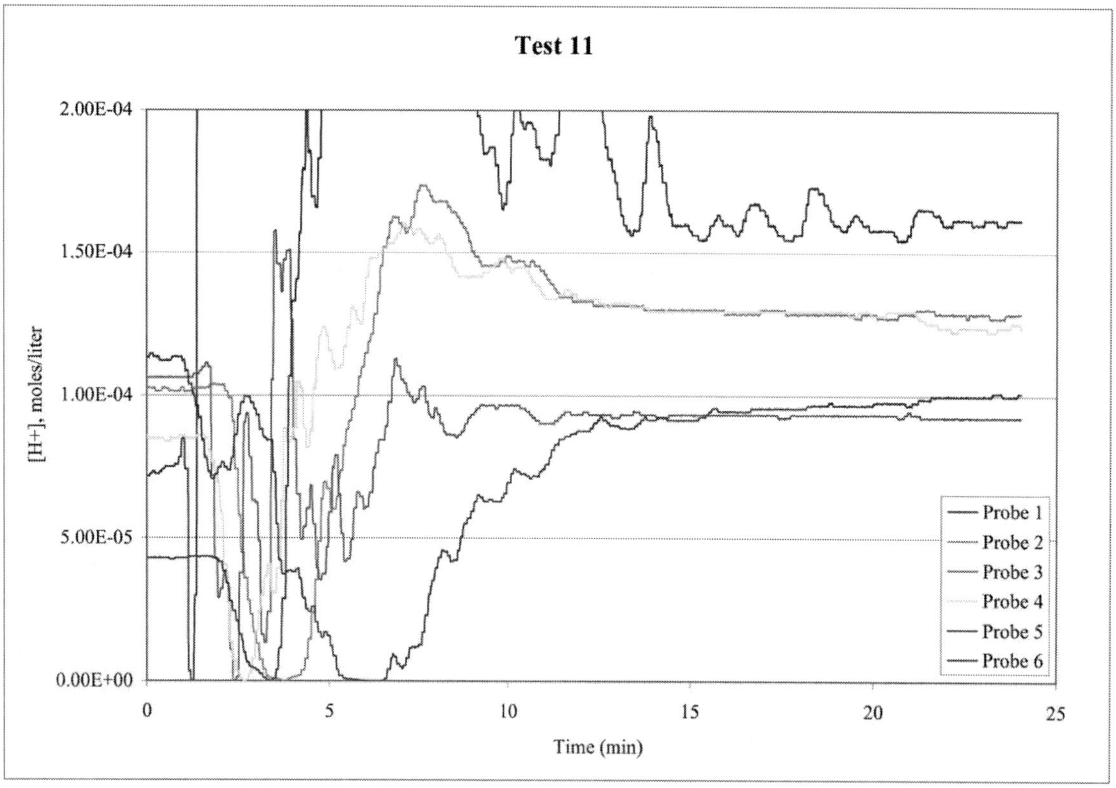

Figure 21: Concentration Data for Analysis Near Equilibrium

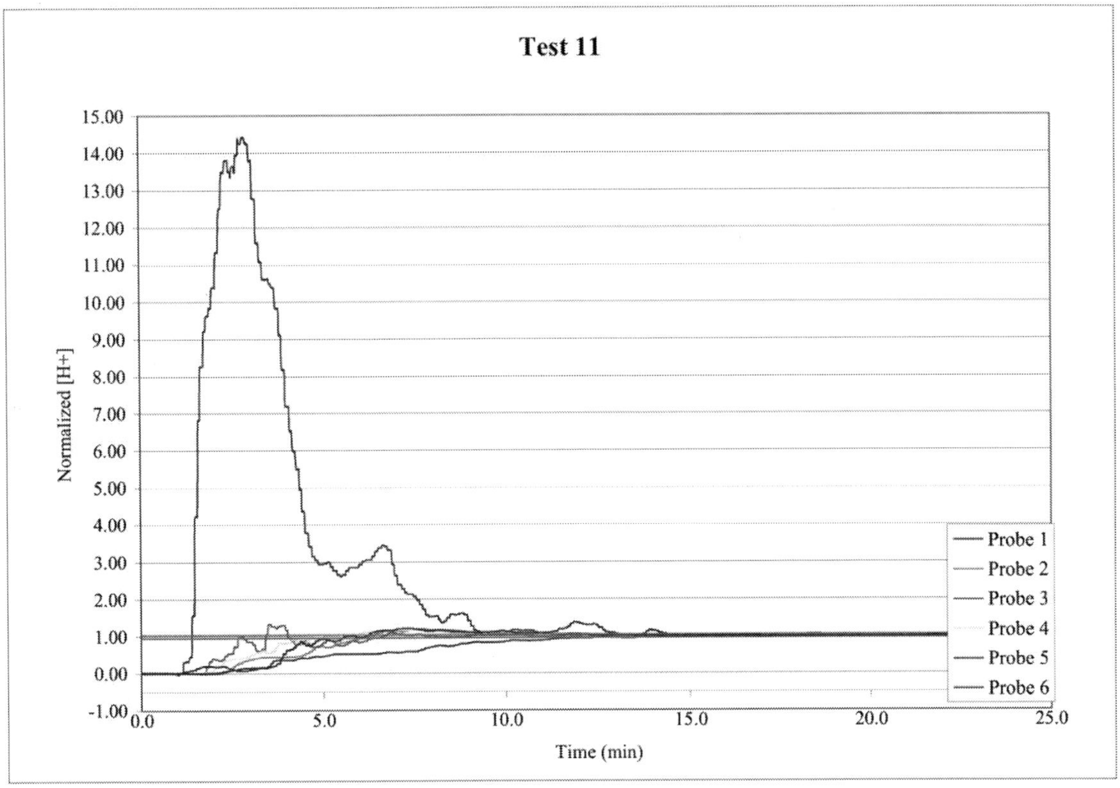

Figure 22: Normalized Concentration Data for Analysis

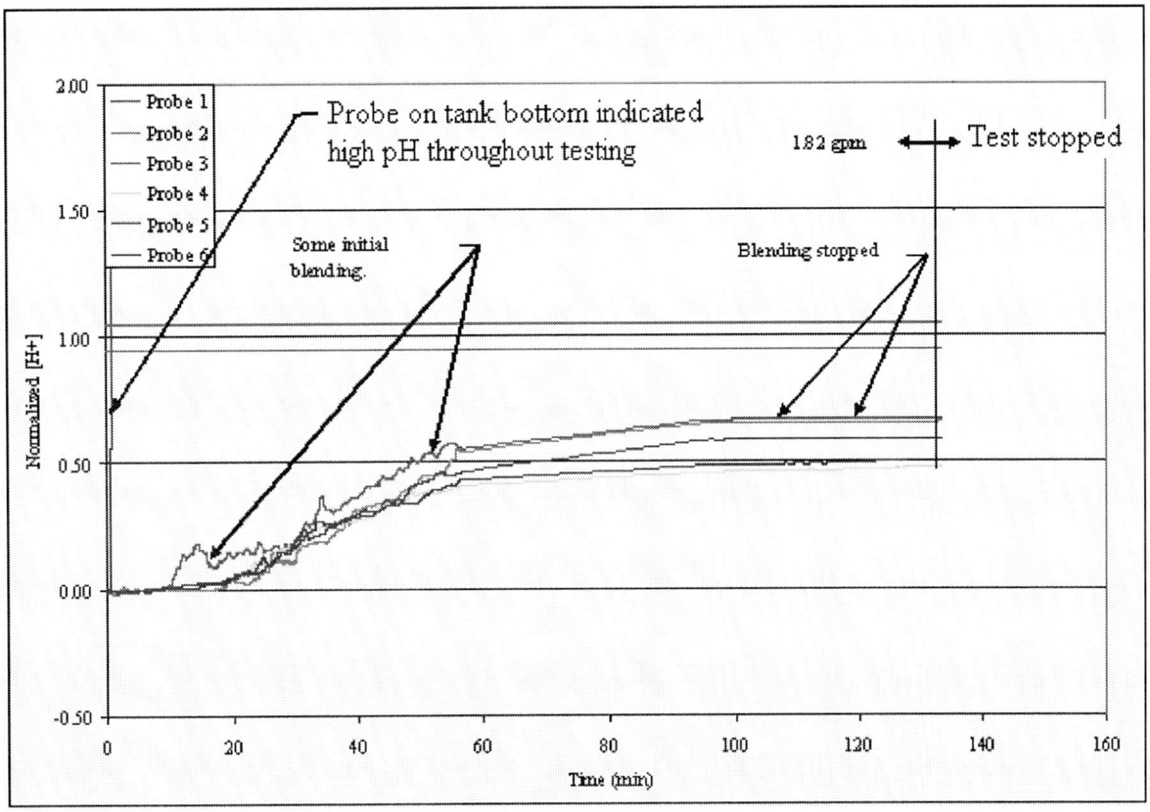

Figure 23: Inadequate Blending (Test 22)

pH Uncertainty

Uncertainties of *pH* measurements are affected by instrument errors and buffering, but are little affected by diffusion. Tests were performed to evaluate diffusion, by adding acid or base to the hopper, and measuring the time for the tank to blend. The tank did not blend in 3-1/2 hours, which is considerably longer than any of the tests. Statistical analyses were performed for all test results and instrumentation.

To obtain a 95% confidence for 95% blending throughout the blending process, the instrument errors were shown to be as large as ± 50.6%. However, instrument errors were shown to be as low as +/- 0.5% near equilibrium. Essentially, data sets similar to Figure 24 had large instrument errors as the blending process proceeded, but the error significantly reduced as the tank contents approached equilibrium. That is, the normalization technique provided a valid estimate of blending effectiveness near 95% blending. Also observed in the figure, some data scatter continues to exceed the 99% blending criteria, which indicates that 99% blending cannot be accurately measured using *pH* probes currently available.

Similarly, buffering of the solution was shown to affect *pH* results. Carbon dioxide in solution contributed a buffering effect to chemical reactions as carbonates and bicarbonates were formed. The concentrations of each chemical were not determined. However, the hydronium ion concentration was the measured blending parameter, and chemical reactions in solution therefore did not affect blending times. In other words, when the reactions were complete, hydronium ions were evenly distributed, and blending was complete.

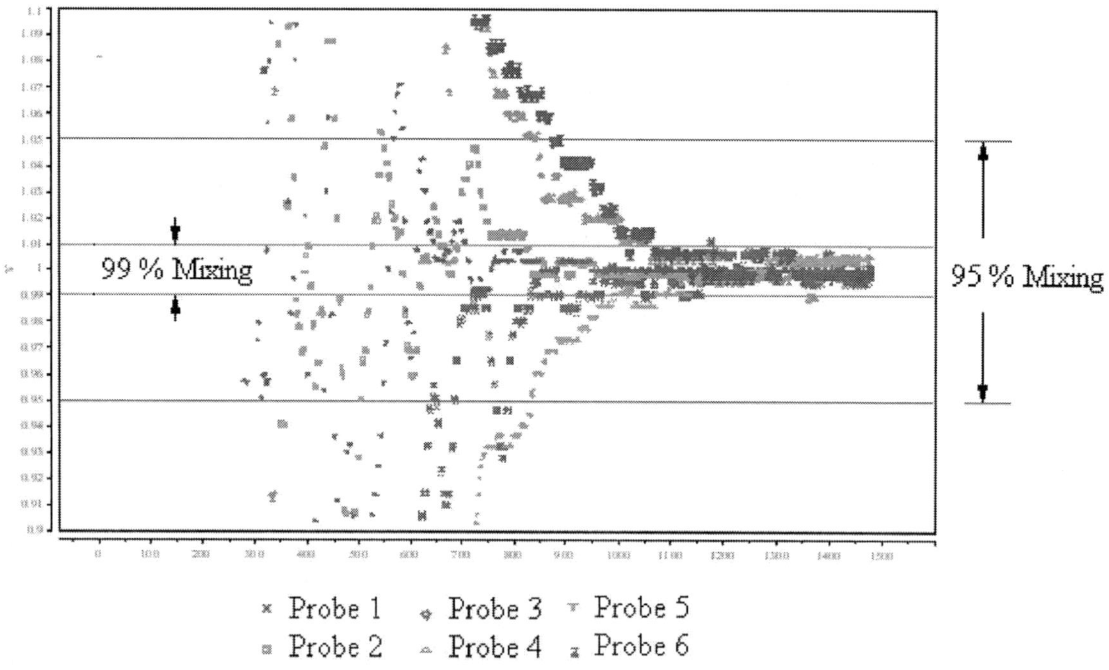

Figure 24: Comparison of 95% Blending to 99% Blending (Test 21)

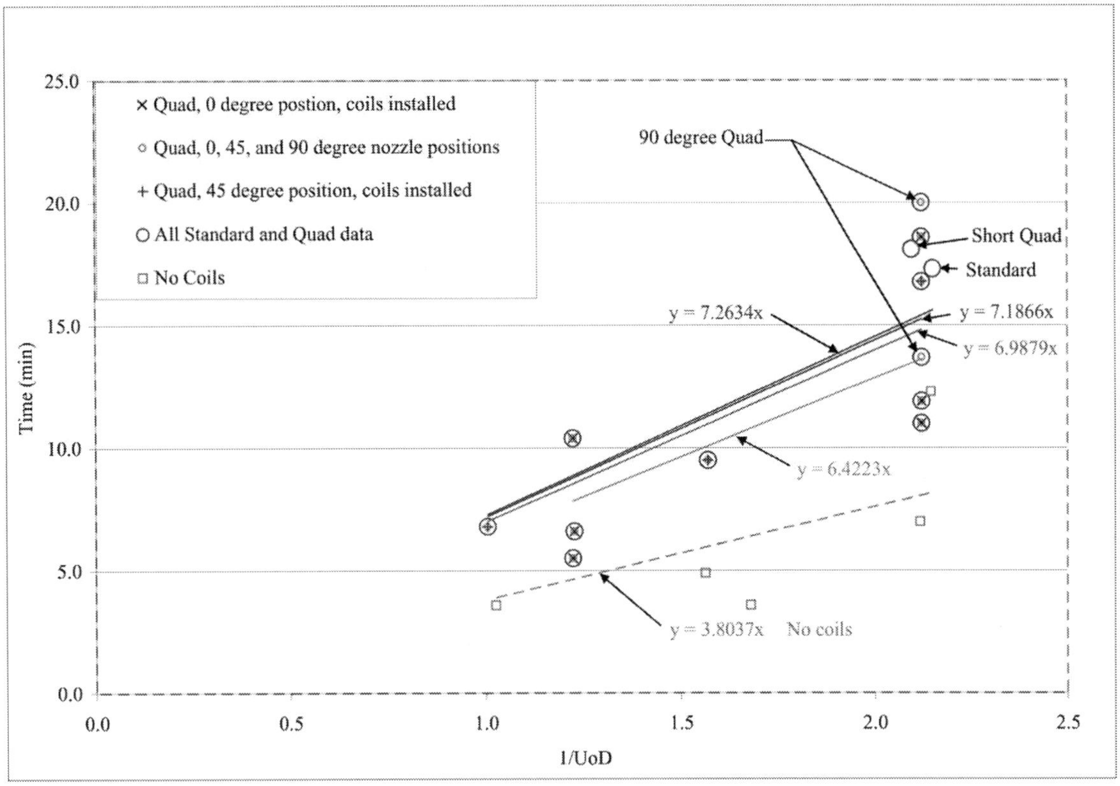

Figure 25: Blending Times for Different Design Parameters (Pilot scale $U_0D \geq 0.47$ feet^2/second)

Blending Time Equation for a Tank Without Coils

Once blending tests were completed, the blending time Equation 1 and Figure 25 were used to find the blending time constant, *C*, such that

$$t(\text{theoretical}) = \frac{3.8037}{U_0 \cdot D} \qquad \text{Equation 10}$$

To determine the constant, Equation 2-1 is rewritten as

$$t(\text{theoretical})(\text{min}) = \frac{C \cdot T^2}{U_0 \cdot D} = \frac{C \cdot \left(\frac{1\_\text{ft}}{12\_\text{in}} \cdot 94\_\text{in}\right)^2}{U_0\left(\frac{\text{ft}}{\text{sec}}\right) \cdot D(\text{ft}) \cdot 60\left(\frac{\text{sec}}{\text{min}}\right)} \qquad \text{Equation 11}$$

Solving Equations 10 and 11, *C* = 3.72. Note that the 3.72 value is close to the 3.0 value recommended by Grenville for a tank without a center column, which was blended by a single jet nozzle normal to the tank wall. This validation for *C* demonstrates the applicability of Equation 1 to the pilot scale tank without coils. Similarly, *C* may be found for any of the curves in Figure 25. Accordingly, the validity of Equation 1 with respect to a tank without coils was confirmed. Statistical analysis of the data in Figure 25 provided a recommendation that the nozzles should be installed parallel to the pump wall, and that insufficient testing was performed to evaluate the effects of nozzle length.

Time Ratios for Tanks With and Without Coils

Test results were compared to the blending equation, using a normalized time ratio, such that

$$\text{Time\_ratio} = \frac{t(\text{measured})}{\dfrac{3.72 \cdot T^2}{U_0 \cdot D}} \qquad \text{Equation 12}$$

where the time ratios and measured blending times, *t(measured)* are listed in Table 1. Figure 26 summarizes test results in a normalized form. As U_0D decreases, non-linearity was observed for both tanks with and without coils as the time ratio increased. Data points at and above U_0D = 0.47 were within the recommended design range. In this range for tanks without coils, the time ratio was near one as expected. Also in this range, time ratios for tanks with coils significantly exceeded one for all data points, and coils therefore had a quantifiable effect on blending.

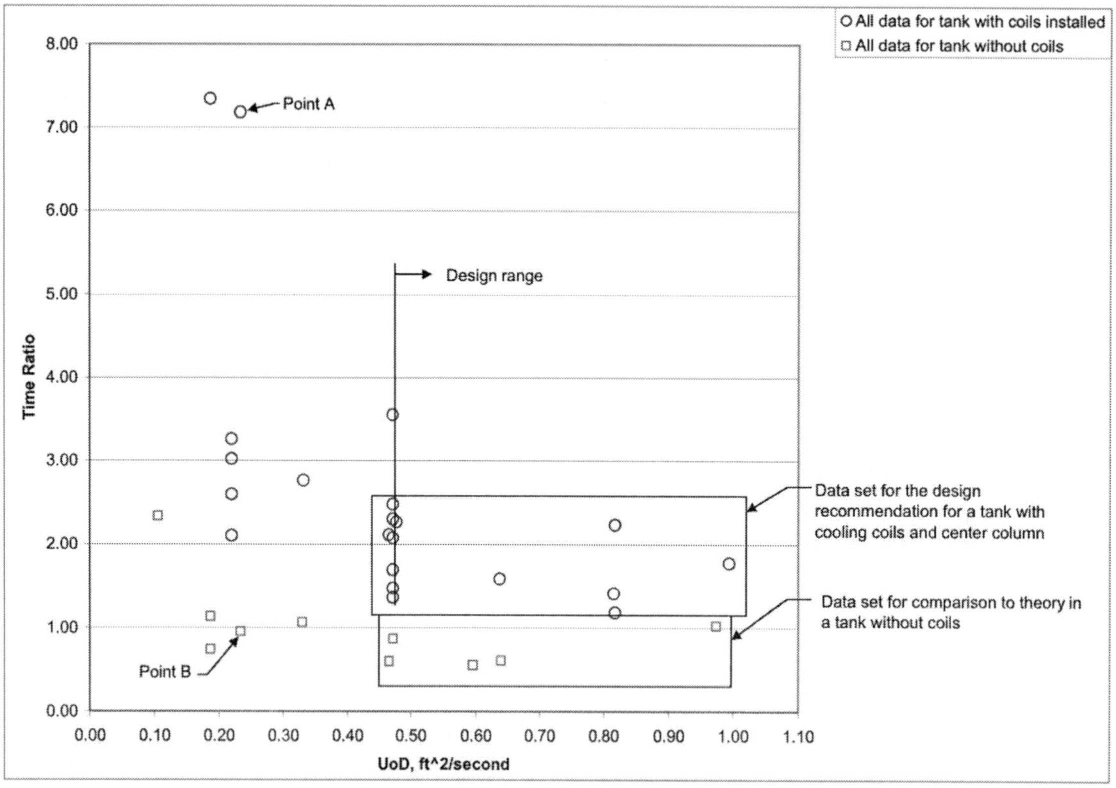

Figure 0-1: Ratio of Pilot Scale Measured Times to Theoretical Times

Conclusions and Recommendations

Pilot scale testing provided new insight into blending in a tank with cooling coils, such that:

- A single pump is adequate to blend the pilot scale tank contents.
- Pump nozzles should be installed parallel to the tank wall.
- The recommended minimum value for U_0D equals 0.47 ft^2 /second for a pilot scale tank filled with water.
- Pilot scale blending times were significantly affected by cooling coil installation, where blending times in a tank with coils were twice the blending times for a tank without coils, when $U_0D > 0.47$ ft^2 /second for the pilot scale tank.
- For a tank with coils when $U_0D < 0.47$ ft^2 /second, the basic fluid mechanics of blending were not understood, and blending times were as much as seven times the blending time for a tank without coils.
- Diffusion had a negligible effect on blending, since the time to blend a tank using diffusion only was considerably longer than the time to blend tank contents using a blending pump.
- Dye testing had apparent blending times at least 1/3 of measured blending times using *pH* probes.
- Nozzle diameters above 0.332 inch diameter and below 0.136 inch diameter were not evaluated.
- Data uncertainty analysis indicated that the 95% commercially used blending criterion provided a valid estimate of full scale blending effectiveness and blending time.

Recommendations for further research include:

- Further testing using a model of the final pump design is recommended.
- Further testing is recommended to quantify blending effects due to higher viscosities. The fluid mechanics and scaling with respect to viscosity and cooling coils in larger diameter tanks are not fully understood.
- Full scale and small scale computational fluid dynamics (CFD) models are recommended for the entire tank with and without coils to validate EDL pilot scale modeling.
- Validations of the CFD models are recommended, using pilot scale blending times and local velocity measurements obtained from small scale testing.

References

[1] Leishear, R. A., Fowley, M. D., Poirier, M. R., "Leishear, R. A., Poirier, M. R., Fowley, M. D., "SDI Blend and Feed Blending Study, Tank 50H Scale Model for Blending Pump Design, Phase 1", SRNL-STI-2010-00054.

[2] Grenville, R., Tilton, J., "Turbulence or Flow as a Predictor of Blend Time in Turbulent Jet Mixed Vessels", Proceedings of the North European Conference on Mixing, 1997.

[3] Grenville, R., Tilton, J., "A New Theory Improves the Correlation of Blend Time Data from Turbulent Jet Mixed Vessels", Transactions of the Institute of Chemical Engineers, Vol. 74, Part A, 1996.

[4] Paul, E. L., Atieno-Oberg, V. A., Kresta, S. M., Eds., "Handbook of Industrial Mixing", Brown, D. A. R., Jones, P. N., Middleton, J. C., "Experimental Methods", Hoboken, John Wiley and Sons, New Jersey, 2004.

(232d) Novel Design for a Magnesium-Enhanced Lime FGD Gypsum Crystallizer Vessel

Chris Hibshman[1], Robert Dowd[1], Wojciech Wyczalkowski[1], Kevin Smith[2], Julia Mercer[2] and Carl Laird[2], (1)Philadelphia Mixing Solutions LTD, Palmyra, PA, (2)Carmeuse Lime and Stone, Pittsburgh, PA

One method of gypsum production in FGD systems is the Magnesium-Enhanced Lime (MEL) external oxidation process. An air sparge grid is the sole means of agitation and solids suspension in this gas-liquid-solids system. The existing design is proven to be effective but is not optimized for agitation, oxygen transfer, or gypsum crystallization. For example, excessive amounts of air and compressor power may be used to achieve agitation than is required for oxidation. This paper presents a novel design of an external gypsum oxidizer/crystallizer, including the addition of a multi-stage mechanical agitator and baffle arrangement to optimize the oxidation and crystallization processes. A pilot scale version of this oxidizer unit has demonstrated a significant reduction in the stoichiometric oxygen requirement and a potential net energy savings of 20-30% over existing MEL FGD installation. Presented data was collected on 0.6 m (24") diameter pilot installation of countercurrent oxidizer/crystallizer consisting of four separate mixing stages. A single full scale version of this new oxidizer design is envisioned to be about 30-feet in diameter, 50-feet tall and capable of processing absorber bleed slurry equivalent to a 700-MW unit burning 7-lb. SO2/MMBtu fuel.

See more of this Session: Mixing in Process Industries
See more of this Group/Topical: North American Mixing Forum

(232e) Energy Efficient Impeller Design for Anoxic Mixing Services

W. Wyczalkowski, Robert Higbee and Binxin Wu, R&D, Philadelphia Mixing Solutions Ltd, Palmyra, PA

Mixing of anoxic basins in Waste Water Treatment Plants require both a relatively uniform suspension of solids and good blending without disturbing the liquid surface. In order to decompose nitrogen compounds, bacteria must be starved of dissolved oxygen which forces the bacteria to consume the chemically bound oxygen within the nutrients. Hydrofoil impellers are typically used for such service. Since energy costs have increased recently, and will increase further in the near future, impeller efficiency is a key issue. A large waste water plant may have between 30-80 anoxic mixers which will consume a significant amount of energy. This presentation discusses an impeller design that is fabricated using composite plastic materials. The choice of composite plastic materials allows one to have 100% control over the blade geometry, unlike typical industrial hydrofoil impellers, which are made by forming flat steel plate. The composite plastic impeller geometry utilizes novel circular rake technology for improved pumping efficiency. Blades are also hyper-skewed in order to avoid the collection of rags. The impeller performance is reviewed and compared to different impeller technologies used for anoxic mixing in terms of blending and solids suspension.

See more of this Session: Mixing in Process Industries
See more of this Group/Topical: North American Mixing Forum

(232f) Turbulent Power Draw During Abnormal Agitator Operation

Kevin J. Myers, University of Dayton, Dayton, OH, Eric E. Janz, Chemineer, Inc., Dayton, OH and **Julian B. Fasano**, Mixer Engineering Company, Troy, OH

On occasion, agitators are operated abnormally, either intentionally or by accident. For example, two of the more common modes of abnormal operation are running in reverse rotation and mounting the impeller upside-down. It is important for both equipment vendors and users to understand the effects of operating abnormally on agitator power draw. In the case of intentional abnormal operation, knowledge of the power draw effect can be used to ensure proper design for all modes of operation that will be encountered. For unintentional abnormal operation, knowledge of the power draw effect can be useful in diagnosing the cause of unexpected power draw. This presentation documents the effects of reverse rotation, upside-down impeller mounting, and other abnormal operating conditions on the turbulent power draw for a number of impellers, including both axial and radial-flow styles.

See more of this Session: Mixing in Process Industries
See more of this Group/Topical: North American Mixing Forum

(409a) Modeling the Mixing of High Concentrations of Bidisperse Cohesive Particles Under Gravity

Lee R. Aarons, Mechanical and Aerospace Engineering, University of Florida, Shalimar, FL

The mixing of granular materials is generally a difficult task. First of all, differences in the constituent particles that would affect their dynamics, such as size, density, and roughness, will generally result in similar particles gathering together [1]. Probably the best-known instance of such behavior is the Brazil nut effect, in which large particles rise to the top of a collection of differently-sized particles that is subjected to vertical shaking. Secondly, small particles (typically less than 100 microns in diameter) tend to agglomerate due to cohesive forces (such as the van der Waals force) exceeding the particle's weight [2]. Naturally, this clustering prevents good mixing. Lastly, it generally takes a good deal of energy to mix solids, as they do not flow as easily as most fluids. A certain amount of stress must be applied to granular materials to get them to flow, and even then, it is not uncommon to find areas where the flow is stagnant or plug-like.

The primary purpose of the present study is to examine the influence of particle cohesion on the homogeneity of mixtures of cohesive particles featuring a large range of particle sizes. As a model problem, we consider discrete element method (DEM) simulations [3] of bidisperse collections of cohesive particles with a diameter ratio of 7:1 undergoing shear flow as a means of mixing. Simulations were performed with different particle cohesive strengths and shear rates. Results have already been presented for simulations in which the particles were not subjected to gravity, so that the Brazil nut (or similar) effect did not occur and cohesion was the only source of particle segregation leading to poor mixing. These simulations provided insight into the influence of cohesion on mixing, but the applicability of the results to real particle mixtures remains to be seen and may be limited. So simulations that feature particles subject to gravity have been performed, and the findings are presented here.

With gravity, the large particles rise to the top and the small particles fall to the bottom as assemblies are sheared. As presented before, when the large and small particles were initially separated side-by-side, layers of large particles reached the top before significant mixing could be achieved in the horizontal direction, and so "reasonably good mixing" could never be achieved. As such, full attention has been given to simulations with initial configurations featuring all of small particles on top of the large particles.

The homogeneity of a given mixture has been measured using the total solid volume fraction, the average size of clusters of small particles [4], and the variance in the concentration of the different sizes of particles. Of those statistics, results suggest that the total volume fraction serves as the best measure of homogeneity. As the large particles spread from the bottom to the top, the mixture naturally becomes more homogeneous, and before the large particles reach the very top, the volume fraction reaches its maximum. As shearing proceeds further, some of the large particles rise out of the mixture and the volume fraction drops. As one may expect, increasing the shear rate results in the peak volume fraction being reached faster. However, the peak volume fraction itself has a complex dependence on shear rate. Also, the maximum volume fraction grows as particle cohesion is reduced, but the rate at which it is reached has a complex dependence on the strength of particle cohesion.

[1] G. Plantard, H. Saadaoui, P. Snabre, B. Pouligny, Surface-roughness-driven segregation in a granular slurry under shear, Europhys. Lett. 75 (2) (2006) 335-341.

[2] J.P.K. Seville, C.D. Willett, P.C. Knight, Interparticle forces in fluidisation: a review, Powder Technol. 113 (2000) 261-268.

[3] P.A. Cundall, O.D.L. Strack, A discrete numerical model for granular assemblies, Geotechníque 29 (1) (1979) 47-65.

[4] S. Gallier, A Stochastic Pocket Model for Aluminum Agglomeration in Solid Propellants, Propellants Explos. Pyrotech. 34 (2009) 97-105.

See more of this Session: Measurement, Monitoring and Characterization Methods for Particulate Systems Research in Powder and Granular Mixing
See more of this Group/Topical: Particle Technology Forum

(409b) Periodic Section Modeling of Convective Continuous Powder Mixing Processes

Yijie Gao, Chem and BioChem Engineering, Rutgers University, Piscataway, NJ, Fernando Muzzio, Chemical and Biochemical Engineering, Rutgers University, Piscataway, NJ and Marianthi Ierapetritou, Chemical and BioChemical Engineering, Rutgers University, Piscataway, NJ

Continuous powder mixing is considered as an efficient alternative to batch mixing in recent years. While modeling work of the axial mixing process on attenuating feeding fluctuations is well developed [1-4], the transverse mixing process on decreasing local heterogeneity [5, 6] is not yet clarified. This hinders standard procedures to be established in design and control of continuous mixing process.

In order to solve this problem, this work aims to develop a general model to describe the transverse mixing process in a convective continuous mixer. The main idea is that continuous mixing can be considered as a combination of powder flow and mixing processes. Powder flow is characterized by the residence time distribution (RTD), whereas powder mixing can be described by batch-like mixing process simulated in one periodic section of the whole mixer. By characterizing the two processes separately, the model can predict the number of sections required to achieve desired homogeneity, as well as the required feed rate to maintain the desired fill level.

Based on this theoretical modeling, convective continuous mixing and batch mixing in the corresponding periodic section are simulated using the Discrete Element Method (DEM) [7, 8]. Segregating and non-segregating mixing cases are simulated to investigate the difference of mixer performance. The results suggest a possible link between continuous and batch mixing in similar geometries, which indicates the potential of this modeling in quantitative design and control of continuous mixing processes. The model is also able to account for the difference of continuous mixing performance due to particle segregation.

As described above, information of the batch-like mixing process is required in successful application of the periodic section modeling. However, due to the net forward flux in the section, this information is not directly achievable experimentally. Thus, the similarity of mixing efficiency between periodic section and the corresponding batch mixer with similar geometries is investigated. Also by means of the DEM, this work examines how this similarity is affected by three design factors: boundary condition of the batch mixer, backward flux in the periodic section and blade location on the shaft. The flow and mixing are quantified using the velocity streamline in a rotating frame of reference, the axial fill level distribution, and the relative standard deviation (RSD) as the mixing index. Simulation results show that satisfactory similarity is obtained when the blades locate on different sites of the shaft, which suggests the possibility to experimentally estimate periodic section mixing by performing corresponding batch mixer. Conversely, similarity is poor when batch mixer is designed with close-end, which mixes much faster than the corresponding periodic section. The results also indicate considerable improvement of the mixing similarity due to the existence of backward flux in the periodic section.

Reference

1. Danckwerts, P.V., *Continuous flow systems : Distribution of residence times.* Chemical Engineering Science, 1953. **2**(1): 1-13.

2. Williams, J.C. and R. Richardson, *The continuous mixing of segregating particles.* Powder Technology, 1982. **33**(1): 5-16.

3. Ralf Weineköter, L.R., *Continuous Mixing of Fine Particles.* Particle and Particle Systems Characterization, 1995. **12**(1): 46-53.

4. Oliver, J.S.I.V., R.G. David, and E.P. John, *Temporal mixing.* AIChE Journal, 2006. **52**(5): 1780-1789.

5. Laurent, B.F.C. and J. Bridgwater, *Influence of agitator design on powder flow.* Chemical Engineering Science, 2002. **57**(18): 3781-3793.

6. Pernenkil, L. and C.L. Cooney, *A review on the continuous blending of powders.* Chemical Engineering Science, 2006. **61**(2): 720-742.

7. Mindlin, R.D., *Compliance of elastic bodies in contact.* Journal of Applied Mechanics, 1949. **71**: 259¨C268.

8. Sarkar, A. and C.R. Wassgren, *Simulation of a continuous granular mixer: Effect of operating conditions on flow and mixing.* Chemical Engineering Science, 2009. **64**(11): 2672-2682.

See more of this Session: Measurement, Monitoring and Characterization Methods for Particulate Systems Research in Powder and Granular Mixing
See more of this Group/Topical: Particle Technology Forum

(409c) On-Line Measurement and Prediction of the Homogeneity of Binary Powder Mixtures in Continuous Mixing During Transitory Regimes

Chawki Ammarcha[1], Cendrine Gatumel[2], Michel Cabassud[3], Jean-Louis Dirion[2] and **Henri Berthiaux**[4], (1)Rapsodee, Ecole des Mines Albi, Albi, France, (2)Research Centre RAPSODEE, Ecole des Mines d'Albi Carmaux, Albi, France, (3)Lgc, ENSIACET, Toulouse, France, (4)Research centre RAPSODEE - UMR CNRS 2392, Ecole des Mines d'Albi Carmaux, Albi, France

The mixing of solids is a crucial unit operation in many industries like the foodstuffs, cosmetics, detergents, powdered metals and drugs. This operation is commonly defined as the blending of different powder ingredients in order to obtain a uniform distribution of at least one key ingredient in the mixture. Unlike for the case of a liquid mixture, the homogeneity of a solid mixture cannot be realized spontaneously and the maximum uniformity of the final product must be measured and assessed. The need to control the blending process is therefore the greatest challenge faced today by the powder mixing technology. In these recent years, on-line and real-time methods for assessing the homogeneity of mixtures homogeneity, such as image analysis, and spectroscopic methods, are experiencing a flourishing development. It may also serve to a better understanding of how the mixing of different particulate flows occurs, and how the homogeneity changes with process variables, with essential aim to control the process.

There are numbers of factors influencing the quality of a final blend like physical characteristics the ingredients, chemical compositions, blending equipment, blending methods and finally process variables or conditions. In the present paper, we will concentrate on studying the effect of process variables variations in the homogeneity of binary mixtures in continuous mixing. In particular, we will focus on the effect of step-like variations in rotational speed of the stirrer, inducing a transitory regime in hold-up weight, flowrate and homogeneity. The apparatus used in this work is a pilot scale commercial mixer Gericke GCM500, for which steady-state operation, pharmaceutical mixing and Markov chain flow modelling has been reported earlier by our reserach team. The particles considered are two sieved fractions of semolina, one of them being coloured in black after iodine adsorption. A linear CCD camera is placed at the outlet of the process and measures the content in any of the two ingredients. The scale of scrutiny can be adjusted and mixture homogeneity can be calculated for this scale in real time. Changes of homogeneity, due to mixer's starting period, mixer's emptying or step-like changes in stirrer's rotational speed are recorded by image analysis. A non-homogeneous Markov chain model developped in a previous work allows the prediction of the homogeneity changes during these transitory regimes. The spatial distribution of the hold up weight inside the vessel for each component is also determined in the different operating conditions. It demonstrate the existence of a light particle segregation phenomena that takes place in the first part of the mixer. The model can also predict this effect and may serve as a good process control tool.

(409d) Effect of Feed Frame Design and Operating Parameters On Powder Residence Time Distribution (RTD)

Rafael Mendez-Roman[1], Luis Guillermo Obregon-Quinones[2], Carlos Velazquez[3], Fernando J. Muzzio[4] and Miguel Florian[1], (1)Chemical Engineering, Engineering Research Center for Organic Particulate System, University of Puerto Rico Mayaguez, Mayaguez, PR, (2)Chemical Engineering, University of Puerto Rico, Mayagüez Campus, Mayagüez, PR, (3)Chemical Engineering, ERC for Structured Organic Particulate Systems, University of Puerto Rico, Mayaguez, PR, (4)Chemical and Biochemical Engineering, Rutgers University, Piscataway, NJ

Powder compression in rotary tablet presses is used as a major step in the manufacture of pharmaceuticals, cosmetics, catalysts, personal care products, military ordnance, etc. The value of all products manufactured by compression operations is enormous; just for pharmaceuticals, more than 70% of all pharmaceutical products sold worldwide are tablets, having an estimated market value in excess of $300 B/yr. In all of these manufacturing processes, consistent filling of the tablet press die with a uniform weight of powder is often both critical to quality and rate-limiting for the entire process, and therefore, uniform die filling is a crucial control variable. The amount of powder entering the die prior to compression determines the weight of the compressed product unit, and for pharmaceutical products, the overall drug content of individual tablets. Powder flowability is affected both by powder material properties and by the equipment used for processing the material. In powder compaction applications, consistent flow of powders into the small die of a tablet press is key to producing tablets with consistent properties. Manufacturers often use mechanical devices, called "feed frames" or "feed shoes" to maintain an adequate supply of powder to fill the dies of the tablet presses. The scope of this project includes the development of effective methods and equipment for continuous feeding of pharmaceutical powders as well as to investigate processing conditions and their powder effects. Through the use of experimental results and DEM simulation, powder phenomena and feeding performance for the two different feed frames (the Manesty Beta Press and the Fette 3090) with different production scale will be investigated. The experimental part includes the residence time distribution (RTD) measurements, the RTD correlation with the shear applied and the RTD and shear-strain effect on powder property changes for un-lubricated and lubricated materials with Acetaminophen or Naproxen Sodium. The experimental data were used to validate the simulation results using the RTD function. After validated the simulation methodology the result were used to understand more fundamentally the powder phenomena inside the feed frame and the contribution of each exit point to the amount of powder in dies.

See more of this Session: Measurement, Monitoring and Characterization Methods for Particulate Systems Research in Powder and Granular Mixing
See more of this Group/Topical: Particle Technology Forum

(409e) Powder Blending Kinetics Using a Controlled Cohesion Apparatus and Its Application

Weixian Shi[1], Dana Hajedemos[2] and Omar Sprockel[1], (1)Bristol-Myers Squibb Company, New Brunswick, NJ, (2)Department of Chemical Engineering, Lehigh University, Bethlehem, PA

The current study presents a novel method to characterize the shear energy of a powder blending system by quantitatively manipulating the cohesive interaction of a minor ingredient to be homogeneously mixed. The blending kinetics was studied using the concept of reaction kinetics (Arrhenius equation). The characterized system can be used to quantify the cohesive energy of a minor ingredient following the methodology presented in the current investigation. In a typical low-shear mixing apparatus, such as bin blenders, the maximum homogeneity of powder mixture and the rate to achieve the homogeneity are important factors to determine whether a low shear mechanism will suffice, which depends on the cohesiveness of the ingredient of interest and the adhesive interaction of that ingredient with other components in the mixture. Applying the concept of reversible chemical reactions, the maximum homogeneity is analogous to the concentration of product at the equilibrium status of a reaction, while the rate of mixing is analogous to the rate of the forward reaction. The current investigation focuses on the determination of the rate constant of mixing. During the initial period of blending, the mixing step (forward reaction) is dominant compared to the demixing step (reverse reaction). This period is thus considered irreversible. The Arrhenius equation was used to estimate the kinetic rate constant for mixing, where the resistance (similar to activation energy of a chemical reaction) originates from the cohesive energy of the component of interest and the driving force (similar to the thermal term RT for a chemical reaction) is exerted by the shear energy due to powder movement. In a typical reaction kinetic experiment, the initial reaction rates at various temperatures are measured and fit against temperature and the activation energy is calculated using the Arrhenius equation. In this study, a magnetic field was used to control the cohesive interaction (activation energy) of iron powder (to mimic the active pharmaceutical ingredients) to characterize the shear energy of a typical pharmaceutical blend in a bottle blender. The use of a magnetic field allows changing only the cohesive interaction between iron particles without inducing other changes in the properties of the iron powder, such as particle size and surface area, which could alter the adhesive interaction between iron powder and other non-magnetized components. Since all other ingredients are non-magnetizable and the loading of iron is low, the shear energy is considered constant under various magnetic fields. An exponential decay relationship is assumed for the mixing kinetics, which has been commonly used since it was suggested 50 years ago by Rose H.E. [1]. The relative standard deviation around the expected concentration (RSDE) of iron is used to represent the homogeneity rather than the normally used standard deviation around the observed mean. The following figures show the model predicted blending homogeneity curve vs. the number of revolutions (in lieu of time). A good match of experimental kinetics vs. the model under various magnetic fields is observed. This model demonstrates that an Arrhenius type relationship is valid for the mixing period and the magnetic attraction (Eind) is the dominant source of cohesive energy. The shear energy of the system was further derived from the Arrhenius equation. [1] Rose, H.E., 1959, A suggested equation relating to the mixing of powders and its application to the study of the performance of certain type of machine; Transactions of the Institution of Chemical Engineers

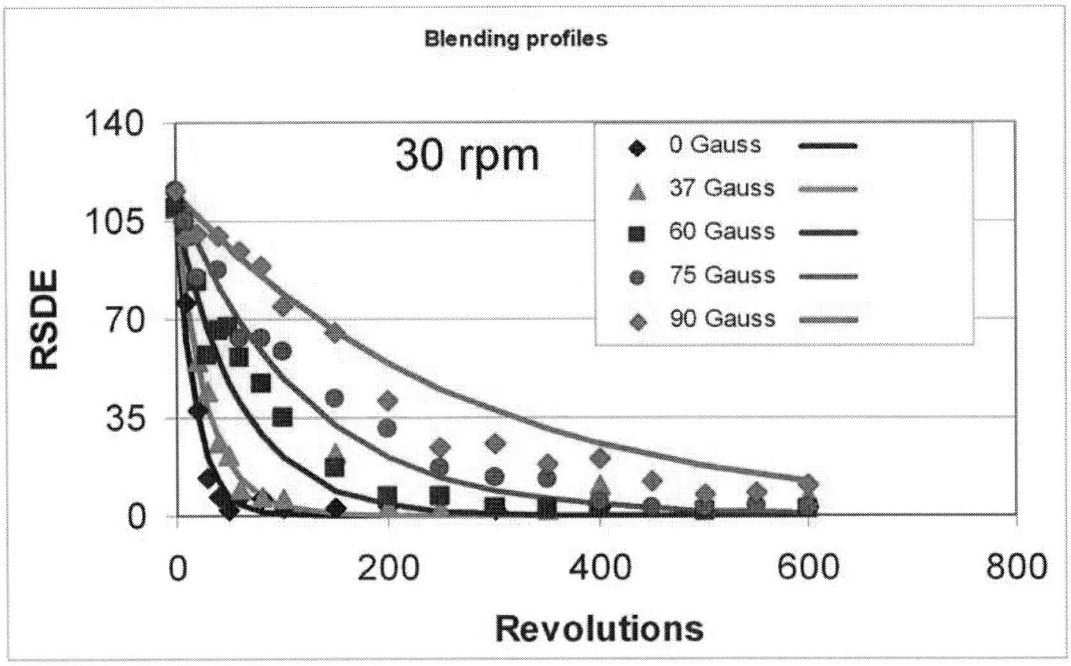
Blending profiles
30 rpm
0 Gauss
37 Gauss
60 Gauss
75 Gauss
90 Gauss
RSDE
140
105
70
35
0
0
200
400
600
800
Revolutions

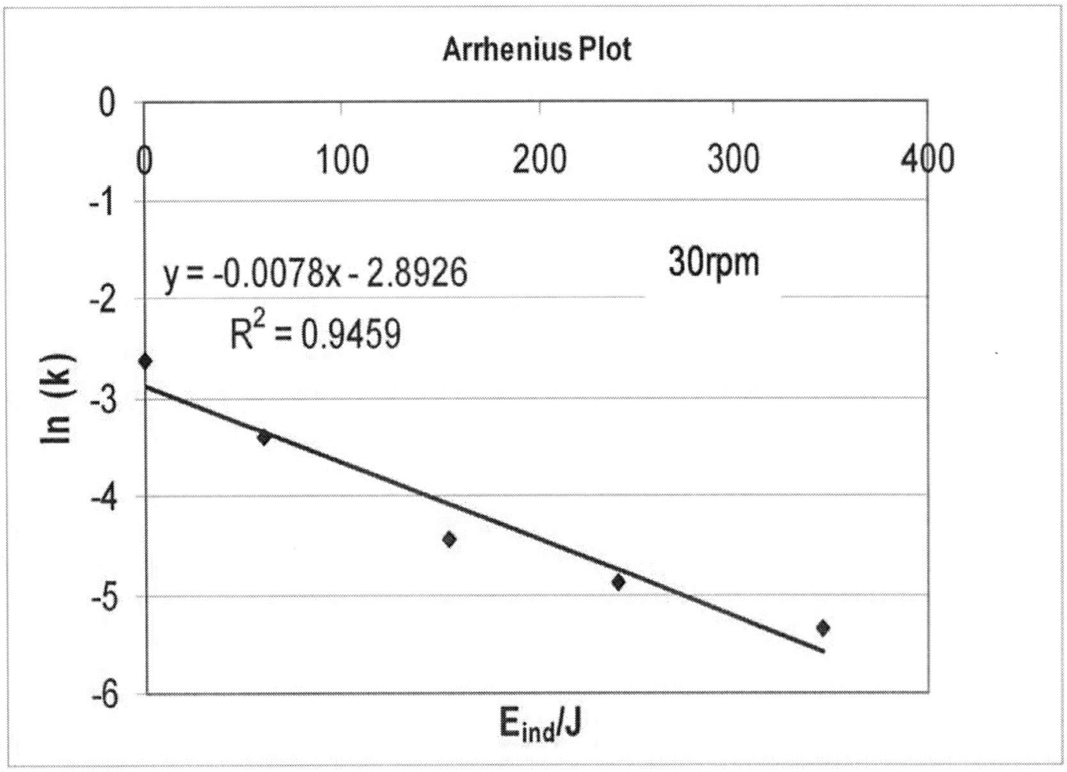
Arrhenius Plot
y = -0.0078x - 2.8926
R^2 = 0.9459
30rpm
ln (k)
0
-1
-2
-3
-4
-5
-6
0
100
200
300
400
E_ind/J

(422a) Proposed Method for Correlation of Reaction Yield On Suspended Catalysts

Gary K. Patterson, Department of Chemical and Biological Engineering, Missouri University of Science and Technology, Rolla, MO

The yield of a complex chemical reaction (two or more reactions in parallel or series) occurring on a suspended catalyst in stirred liquid phase depends on the relative rates of the chemical reactions and the mass transfer rates of the reactants to the catalyst and of the products away from the catalyst If the blending rate of a fed reactant into a pre-existing mixture of a resident reactant and the suspended catalyst is much faster than interphase mass transfer onto the catalyst and the chemical reactions in the liquid phase are very slow relative to those on the catalyst, then a Damkoehler number based on the characteristic times for mass transfer and chemical reactions should serve to correlate yield of the primary reactant. Numerical experiments have shown that for a fed-batch reactor the feed time of the fed reactant is a more important variable than the characteristic reaction rate, probably because in such a system feed rate determines the actual rate of reaction on the catalyst. Higher feed rates induce higher solution reactant concentrations, resulting in higher mass transfer rates and higher surface concentrations of the fed reactant. A higher reaction rate on the catalyst results. Numerical results show the likely effectiveness of such a correlation method, but actual experimental data are needed to confirm the results. Thresholds of blending rate and mass transfer rate relative to chemical reaction rate are explored.

See more of this Session: Progress in Mixer and Reactor Design
See more of this Group/Topical: North American Mixing Forum

(422b) Can This Process Be Saved? A Search for Understanding Using the Bourne Protocol and Advanced Process Development Tools

Cheryl Teich, Core R&D/ Process Chemistry & Development, The Dow Chemical Company, Spring House, PA, Aaron Sarafinas, Core R&D / Engineering Sciences, The Dow Chemical Company, Spring House, PA and Paul Morton, Core R&D / Analytical Sciences, The Dow Chemical Company, Spring House, PA

An interplant process transfer offered an excellent opportunity to effect a change that promised a significant reduction in the cycle time of a specialty solid product. The new reactor delivered the necessary residence time for the chemistry and bulk mixing, but neglected other controlling effects. The end result was a smaller-sized product that impacted not only this product, but also the operations of other processes across the entire plant. The Plant operating staff had devised a fix, but its associated cost penalty was unsustainable for the long-term.

While initial discussion with the Plant centered on "fixing" the process' downstream train, the focus moved quickly to the reactor and consideration of the most economical design for the specialty product. Based on the protocol Bourne presented in "Mixing and the Selectivity of Chemical Reactions" (Org. Proc. Res. Dev., 2003, 7 (4), pp 471–508), we ran a series of experiments in the METTLER TOLEDO RC1e™ reaction calorimeter equipped with a METTLER TOLEDO FBRM® in-situ particle characterization system. Results of those experiments, combined with solid-liquid equilibrium data modeled using an OLI electrolyte simulation, led to an understanding of why the new reactor made small particles, how the Plant "fix" mediated the problem, and finally, to identification of an optimal reactor design and operation conditions that would produce low-cost, right-sized, product.

See more of this Session: Progress in Mixer and Reactor Design
See more of this Group/Topical: North American Mixing Forum

(422c) Digital Imaging, and Subsequent Quantification, of Mixing in Bioreactors

Claudia Berdugo[1], Adeline Sadeli[2], Robert S. Brodkey[3], Jeffrey J. Chalmers[1] and Yang Zhao[2], (1) Chemical and Biomolecular Engineering, The Ohio State University, Columbus, OH, (2)The Ohio State University, Columbus, OH, (3)The Department of Chemical and Bimolecular Engineering, The Ohio State University, Columbus, OH

Traditional mixing studies use single point measurements at a specific location to measure mixing. While this approach presents a single location information of the mixing conditions it does not represent the entire mixing performance in the vessel. To overcome these difficulties protocols have been developed that include more than one probe in order to obtain more representative measurements that allow a better estimate of mixing performance. Those studies have shown that measurements in different positions are notably different, reinforcing the need for data from multiple locations and further introducing the concept that one has not investigated enough locations. An alternative to a single point location has been the addition of dyes, or other similar type of indicators, that allow visual observations of the mixing in the whole vessel. However, these techniques tend to much more qualitative in nature. In this study we applied a commercial, digital image processing software which allows a "region of interest", ROI, to be identified, and the subsequent distribution in pixel intensity for all the pixels in this ROI to be recorded frame by frame. This technique was used on several vessels; with at least one vessel the ROI represented approximately 40% of the total volume of the vessel. Such large volumes allow a much more rapid evaluation of a range of agitation speed, impeller designs, and energy dissipation.

See more of this Session: Progress in Mixer and Reactor Design
See more of this Group/Topical: North American Mixing Forum

(422d) Characterization of Mixing in Chemical Reactors with Mean Age and Higher Moments

Minye Liu, DuPont Engineering Technology, Wilmington, DE

In this talk, a new theory for characterizing mixing processes in chemical reactors will be discussed. This theory is based on the conservation equations of mean age and higher moments of age in a reactor. These equations are in the same form as N-S equations and can all be solved with the same CFD solvers that are used to solve for the velocity solutions. For a steady continuous flow system, these equations can be solved after the velocity field is obtained. Compare to the current methods of computing residence time distributions by solving time dependent scalar concentration equation or by tracking particles, this new method is not only extremely efficient in using computing resources but also much more accurate in solving for the moments of age inside the reactor and the moments of residence time at the exit. Several examples of steady continuous reactors will be discussed to demonstrate how this new theory can be used to quantitatively characterize mixing performance.

See more of this Session: Progress in Mixer and Reactor Design
See more of this Group/Topical: North American Mixing Forum

(422e) Simplification and Application of An Interfacial Area Concentration Balance by Order Arguments

Richard L. Long, New Mexico State University, Las Cruces, NM

Two approaches have been developed for dealing with two phase systems. In the first approach mass balances for the two phases with no source or exchange terms, a total momentum balance for the two phase mixture, and a gas phase momentum balance with source term models for forces based on a two fluid approach have been developed. In effect, the action of the liquid on the gas phase has been included, but the effect of the gas on the liquid pahse hase been neglected. The results obtained are very good in modeling gas-liquid flow in a vertical tube. One significant limitation is that in the total momentum balance used in this approach, the gas phase momentum is neglected, so that the appoximation is only valid over a constrained range of two phase flows. Clearly, the dispersed phase momentum being neglected would introduce much larger errors at a given phase fraction if both fluids were liquids. Thus, this aproch is limited to gas-liquid systems.

In the second approach (actually developed first) mass balances with no source or exchange terms were combined with a momentum balance for each fluid phase and with exchange terms between the phases and an interfacial energy balnce with a source term from the continuous phase mechanical energy balance was developed. Computations with this apporach have continued to show problems. Because of these problems, an effort was made to simplify the approach by simplifying the source term in the interfaial energy balance with order arguments and a simple macro model for the two fluid friction factor. Results are encouraging.

See more of this Session: Progress in Mixer and Reactor Design
See more of this Group/Topical: North American Mixing Forum

(476a) Novel Stirred High-Throughput Mini-Bioreactors for Bioprocess Development: Evaluation of Scale-Down From a Typical Bench-Scale Bioreactor

Bhargavi Kondragunta[1], Joao Silva[1], Kurt A. Brorson[2], Antonio R. Moreira[3] and Govind Rao[1], (1) Chemical and Biochemical Engineering, UMBC, Germantown, MD, (2)Center for Drug Evaluation and Research (CDER), Food and Drug Administration (FDA), Silver Spring, MD, (3)Chemical and Biochemical Engineering, Center for advanced Sensor Technology. University of Maryland, Baltimore County (UMBC), Baltimore, MD

High-throughput bioreactors could offer an economical alternative to bench-scale systems for Bioprocess development. They could also bridge the gap between un-instrumented culture devices such as shake-flasks, and instrumented bench-scale systems thus lowering the development time at the bench-scale. Bench-scale bioreactors that are the mainstay in cell culture process development are widely accepted as scale-down models of large-scale impeller driven bioreactors. The novel high-throughput mini-bioreactors (HTBRs) not only have the advantage of non-invasive, disposable optical sensing and control for %DO and pH, but also have overhead stirring that sets them apart from the existing high-throughput systems that mainly rely on magnetic stirring or shaking. However, the new HTBR system needs to be validated as a scale-down model of the existing bench-scale bioreactors to be certain that HTBR data is transferable to other bioreactor scales. Generally, physical parameters such as k_La, mixing time, P/V are used for scaling. However, it is recognized that engineering challenges exist to obtain complete comparability between the two scales, as depicted by cell growth and product quality which are the true measures of scalability. This scale-down study aims to study the effect of HTBR and conventional reactor design as well as environmental parameters on cell culture process scale-down, including a comparison at the gene level using DNA microarrays. This study is also a first demonstration of mammalian cell culture in new controlled mini-bioreactors. To achieve a 150-fold scale-down from bench-scale to the HTBR, oxygen transfer rate (OTR) was matched. However, complete cell growth and product comparability was not achieved. An initial growth lag as well as lower end product titer was observed. These were surmised to be attributable to oxidative stress and free-radical damage conditions in the HTBRs; and preliminary DNA microarray results corroborated this. Further scale-down evaluation was done by comparing the mixing times between the mini-bioreactor and bench-scale, using the paddle and pitched blade impeller configurations. Based on our findings, if relevant modifications can be made to the HTBRs, scalability can be achieved pending further scale-down studies.

See more of this Session: Mixing and Scale-up/Scale-Down Strategies for the Production of Biopharmaceuticals
See more of this Group/Topical: North American Mixing Forum

(476c) Mixing Uniformity Characterization of 15K Mammalian Cell Culture Bioreactor

Mei Shao[1], David Lindsay[1], Brian Stamper[1], LiYing Yang[1], Robert Torregrossa[1], Bernd Schmidt[2], Darren Hodson[2], Alistair Boyds[2] and Gavin Reynolds[2], (1)MEDIMMUNE LLC, Frederick, MD, (2) ASTRA ZENECA, United Kingdom

The mass transfer uniformity of a 15,000 liter mammalian cell culture bioreactor with modified pitched blade impellers was demonstrated by empirical methods and computational fluid dynamics (CFD) modeling. The gas dispersion, mass transfer and blending performance of the 15,000 L bioreactor were characterized empirically. Two empirical methods were used to evaluate the bioreactor blend times under different agitation speeds. A computational fluid dynamics (CFD) model was developed using ANSYS Fluent. The CFD model was verified by empirical blend time results. The flow pattern and mass distribution in the 15,000 liter bioreactor over a range of impeller operating speeds were characterized with CFD modeling. The results from empirical blend studies and CFD modeling indicated top/bottom blend time uniformity. Also, zonal uniformity of oxygen transfer was demonstrated by gas dispersion and mass transfer characterization studies. The study results and lessons learned will be discussed.

See more of this Session: Mixing and Scale-up/Scale-Down Strategies for the Production of Biopharmaceuticals
See more of this Group/Topical: North American Mixing Forum

(476d) A Scale-Down Study of the Impact of Different Stress Parameters On the Growth and Productivity of GSCHO Cell Culture

William H. Scott[1], Colin R. Thomas[1], Chris J. Hewitt[2], Gareth Lewis[3] and **Alvin W. Nienow**[1], (1) Chemical Engineering, University of Birmingham, Birmingham, United Kingdom, (2)Chemical Engineering Department, Loughborough University, , Loughborough, United Kingdom, (3)Cell Science, MedImmune, Cambridge, United Kingdom

This work aimed to study the effect of potential large-scale bioreactor environmental heterogeneities during fed-batch culture on the performance of an industrially-relevant GSCHO cell culture, producing an IgG antibody. Heterogeneity was created by applying, for the first time in animal cell culture, a two-compartment scale-down model, using a combination of a well-mixed stirred tank reactor (STR) and plug flow reactor (PFR). Feeding of glucose substrate and alkali for pH control to the PFR was analogous to feeding them to the liquid surface of a large-scale bioreactor. The flow rate through the PFR was controlled with a peristaltic pump so that the mean residence time in the PFR was equal to typical mixing excursions at the ~ 20 m3 scale. In this way, perturbations created in the PFR were analogous to perturbations created by poor mixing typical of the large-scale. The results obtained were compared with controls in which either just the STR was used; or in which circulation took place but all feeds were introduced directly into the STR. In addition to the standard parameters, antibody titre and quality were measured and flow cytometry was used to indicate cell viability and the mode of cell death. For the latter, viability was monitored by dual staining with Calcein-AM and Propidium Iodide (Calcein-AM/PI) and the mode of death by dual staining with Annexin V conjugated to phycoerythrin and Sytox Green (AV-PE/SG). Typically growth was continued for ~ 20 days and for the STR/PFR runs, this required continuous pumping for this time with a peristaltic pump. Durable neoprene was used in the pump head and the PFR tube was silicone. The results essentially fell into two categories; those without circulation and those with it. In all cases with recirculation whether nutrients and alkali were added into the STR or the PFR, significantly reduced culture duration (by ~ 48 hours) and antibody titre (~ 20% reduction) were found compared to those runs without circulation. All other key process indicators were the same for all cases, with or without recirculation, including death by necrosis. The equivalence of antibody quality in even those cases with greatly reduced viability provided strong evidence for robust antibody production in this cell line. Clearly in this study, it was not possible to conclude anything concerning the impact of bioreactor heterogeneities with this cell line. On the other hand, damage associated with peristaltic pumping has relevance to the many aspects of cell culture processes that require transfer of cells in suspension; for example, inoculation and harvest steps. For large culture volumes, pumping duration may be of significant duration. It is considered that the 'squeezing' motion by which peristaltic pumps cause flow may impose sufficient mechanical stress on the cells to cause the relatively poor performance. It is of course possible that it may be due to long term chemical leaching or other features of the flow loop but with the high biocompatibility of the materials used in this study, this reason is considered to be less likely.

See more of this Session: Mixing and Scale-up/Scale-Down Strategies for the Production of Biopharmaceuticals
See more of this Group/Topical: North American Mixing Forum

Enabling T-flasks to improve bioprocess development: A novel disposable cell culture train from cell thawing to bench-scale.

Jose R. Vallejos[a], Kurt A. Brorson[b], Antonio R. Moreira[a], Gary J. Lye[c], Martina Micheletti[c], Govind Rao[a]*

a Center for advanced Sensor Technology, Department of Chemical and Biochemical Engineering, University of Maryland, Baltimore County, 1000 Hilltop Circle, Baltimore, MD, 21250, United States
b Office of Biotechnology Products, Center for Drug Evaluation and Research, Food and Drug Administration
10903 New Hampshire Ave., Silver Spring MD 20903, United States
c The Advanced Centre for Biochemical Engineering. Department of Biochemical Engineering University College London, Torrington Place, London, WC1E 7JE, U.K.

During the last five years disposable cell culture technology has become increasingly popular for laboratory scale studies and seed culture generation. In particular, T-flasks are disposable low shear devices commonly used for cell culture in both academia and industry settings. However, even this simple single-use technology lack accurate engineering characterization of critical physical parameters such as gas-liquid mass transfer coefficient (k_La), power input per unit volume (P/V), mixing time as well as velocity characteristics. In this work mass transfer and hydrodynamic studies have been carried out in T-flasks, under static, shaken and rocking agitation, using patch-based disposable optical sensors and Particle-Image Velocimetry (PIV), respectively. The accurate measurement of critical parameters like k_La and temporal and spatial variations of shear stresses have provided fundamental insight for the successful scale-up from T-flasks under rocking agitation to conventional stirred bioreactors. A number of parameters will be monitored including dissolved oxygen (DO), pH, cell density, product titer and glucose to lactate molar ratio at every stage of the process from cell thawing to bench-scale bioreactor. A control experiment will also be investigated using static T-flasks. This research study may provide novel insights into the type of physiological stress (e.g. respiration rate, glucose to lactate molar ratio) experienced by cells during scale-up, from thawing to bench-scale cell culture devices. From a chemical engineering and regulatory perspective, this study could show that process understanding in T-flasks and improved scale-up strategy can allow integration of small Process Scouting Devices (PSDs) with subsequent larger scale cell culture systems. Ultimately, this work aims at integrating fundamental engineering insight, disposable sensors and cell culture technology to facilitate the future development of a disposable T-flask based cell culture train for clone selection and early seed expansion stages.

In the first stage of this project, we carried out a mass transfer characterization of T-25, T-75 and T-150 flasks under static, shaking (i.e. orbital) and rocking agitation conditions. We determined the volumetric mass transfer coefficient (k_La) and then compared the values obtained with results obtained in alternative PSDs systems such as spinner flasks. We observed that T-flasks can be operated at similar k_La values as Spinner Flasks thus allowing us a cell culture comparability study between two cell culture trains during scale up. Additionally our results show that k_La values as high as 22.6h^{-1} were obtained in T-25 Flasks when operated under shaking conditions while under rocking agitation the highest $k_L a$ observed decreased to $14.8h^{-1}$. The same trend was observed in the other T-Flask systems. This mass transfer study has some further bioprocess implications. In particular, it shows that in the future T-25 could be equipped with DO and pH sensors in a novel T-25 platform with high throughput capabilities (as in 24 well plate cassettes). Clone selection and media optimization studies at similar $k_L a$ values to those typical of bench-scale bioreactors could also be performed in this novel T-25 platform. Finally, results from our study could help designing a novel scale-up strategy from PSDs (clone selection and early seed expansion stages) to bench-top bioreactors (early process optimization and preclinical studies) using a fed-batch system at similar $k_L a$ values across scales. This novel approach has the potential of avoiding significant differences in product titer and product qualities attributes between clone selection and early process optimization (previous to preclinical studies) that could delay the overall bioprocess timeline.

Scale-down Studies of the Effect of Hydrodynamic Forces on CHO Cells: Implications for Industrial Production Conditions

W. Scott[1], S. Meier[1], R. Kiss[1], A. Amanullah[1], A. Nienow[2]
[1] Pharma Technical Development, Genentech, Inc., USA
[2] University of Birmingham, UK

The effect of hydrodynamic forces on CHO cells used for the production of biopharmaceuticals has been studied by multiple academic and industrial researchers. Such studies have included evaluation of lethal forces, as well as non-lethal forces, for their effect(s) on cell growth and productivity as well as recombinant protein quality attributes. In this work, two different CHO production cell lines were studied in a repetitive defined elongational shear field device, utilizing two types of pumping systems (peristaltic and syringe pumps). Results to date show that the peristaltic pumping action required to drive the flow through the shear device causes more damage than the shear device itself at low flow rates. Growth and productivity are affected, but not protein quality. Syringe pumps, on the other hand, do not appear to show any significant adverse effects under the flow rates studied. Energy dissipation rates through the shear device are compared to a variety of energy dissipation rates in stirred tank bioreactors, the typical production vessel for industrial purposes. Even energy dissipation rates much higher than those required practically at the current cell densities produced in bioreactors do not show any adverse effects on performance or protein quality with one cell line, consistent with previous results showing no significant effects on growth or productivity with multiple cell lines. This work will be summarized in the context of practical conditions in typical stirred tank bioreactors, and the likelihood that such hydrodynamic forces would significantly impact process performance or protein quality attributes.

(476g) Directional Oscillatory Shear Index in Orbiting Petri Dishes

Amlan Chakraborty[1], M. Keith Sharp[2], Venkatakrishna R. Jala[3], Haribabu Bodduluri[3] and **R. Eric Berson**[1], (1)Chemical Engineering, University of Louisville, Louisville, KY, (2)Dept. of Mechanical Engineering, University of Louisville, Louisville, KY, (3)Microbiology & Immunology, University of Louisville, Louisville, KY

Wall shear stress (WSS) is widely accepted as a primary influence affecting characteristics of anchored cells subjected to fluid flow. Common orbital shakers provide a means of simultaneously applying shear stress to cells for tens to hundreds of cases by loading the shaker with multiple dishes. Orbital shakers allow simultaneous cell culture experiments and are widely used throughout the cell culture industry because of their simplicity. The movement of fluid in a dish that derives its motion from an orbiting shaker platform will be oscillatory in nature, somewhat like the pulsing fluid movement in the human vasculature system, with a wave whose peak rotates around the dish at an angular velocity corresponding to the orbital velocity of the dish. However, the complex flow in orbiting dishes is generally not amenable to analytical solution for resolving shear created by the fluid motion except for limited simplified conditions. The only existing quantification of shear in this flow is an equation that estimates a constant scalar value of shear for the entire surface of the dish. In practice, WSS will be oscillatory in both magnitude and direction rather than steady due to the traveling waveform and will vary across the surface of the dish at any instant in time. Computational fluid dynamics (CFD) was, therefore, employed to simulate the motion of fluid in orbiting dishes. A model was created that provides complete spatial and temporal resolution of WSS over the bottom surface of the dish and validated using particle image velocimetry.

An oscillating shear index (OSI) was previously defined for use in the analysis of arterial disease to describe the cyclic departure of the WSS vector from its predominant axial alignment (Ku et al., Atherosclerosis, 5, 293-302,1985). We modified the index for the case of an orbiting petri dish where the fluid motion can be described in terms of radial and tangential components and termed it a directional oscillating shear index (DOSI). While the DOSI does not describe the frequency of oscillation, it does quantify the oscillatory nature between the radial and tangential contributions to the shear stress. The value ranges from 0 that indicates 180o cyclic variation or bidirectional flow to a maximum value of 1 that indicates no cyclic variation between the radial and tangential components or unidirectional flow. Examples are presented that show how endothelial cell proliferation and morphology are affected by shear magnitude and the DOSI, which both vary across the area of the dish, and help explain why cellular responses depend on location in the dish.

See more of this Session: Mixing and Scale-up/Scale-Down Strategies for the Production of Biopharmaceuticals

(522a) Computing Blend Time with Mean Age Distribution in Batch Stirred Tank Reactors

Minye Liu, DuPont Engineering Technology, Wilmington, DE

Blend time is a basic measure for performance efficiency of a stirred tank reactor. In the last two decades, there have been several occasions that CFD was attempted to compute blend time with RANS model but the reported results have been inconsistent. More recently, LES has been used to compute the time dependent flow solution and tracer history. Good results in blend time predictions have been reported but the huge CPU time demanded by the computation is still an obstacle for the method to be used in industrial reactor design. In this talk, a new method for computing blend time in a stirred tank reactor will be discussed. This method is based on the mean age distribution that can be obtained by solving a conservation equation of passive scalar. Unlike the current method of tracking time history of a passive tracer by solving the time dependent concentration equation, the new method solves only a steady scalar equation for mean age distribution in the reactor. Therefore, the method is extremely efficient numerically. The CPU time consumed by this new method is orders of magnitude shorter than that of the above mentioned current methods. This method makes it possible to predict blend time of an industrial stirred tank reactor on a desk top computer. Different types of impellers will be studied with this new method and the predicted blend time will be compared with experimental results and correlations available in literature. The use of the method for scale-up of stirred tank reactors will also be discussed.

See more of this Session: CFD and Mixing Effects in Chemical Reaction Engineering
See more of this Group/Topical: North American Mixing Forum

(522b) A Comparison of One Dimensional Turbulence (ODT) and Direct Numerical Simulation (DNS) of Non-Premixed Flames with Extinction

David O. Lignell and Devin S. Rappleye, Chemical Engineering, Brigham Young University, Provo, UT

The modeling and simulation of turbulent reacting flows with strong finite rate chemical kinetic effects is a significant challenge. Flows of this type are common in combustion involving pollutant formation (e. g., NOx, and soot), and flame extinction and reignition phenomena. This complexity arises from the wide range of turbulent length and timescales, which cannot be fully resolved in high Reynolds number simulations, along with the highly nonlinear nature of chemical reaction rates, the presence of complex flame and flow structures, and multicomponent transport effects. In order to develop more advanced and capable combustion models, two important and related types of information are required: (1) fundamental information of physcal processes of transport-chemistry interactions in turbulent flows; and (2), data for model validation. Direct numerical simulation (DNS) is an unparalleled tool for probing the details of reacting flows, and is continually improving in its physical realism and range of accessible flows as computational resources expand. Modern DNS, however, require millions of CPU-hours for detailed combustion simulations at low-to-moderate Reynolds number. These simulations are typically computed in isolation, or perhaps including a few parametric extensions to a baseline configuration. This high cost precludes detailed investigation of parameter sensitivity, or problem scoping. The one-dimensional turbulence (ODT) model is a simulation approach in which high computational costs are mitigated through solution in only a single dimension, while turbulent advection is modeled stochastically by spatial maps whose size, location, and frequency are implemented using turbulent scaling laws based on the local instantaneous velocity field. These maps increase scalar gradients in a manner consistent with correponding turbulent advection. Computations are resolved in physical space with reactive-diffusive evolution of scalar fields. Hence, the ODT model may provide useful information for model validation and parametric investigation of turbulent flows at a cost orders of magnitude lower than that of DNS. While it is not implied that ODT is a replacement for DNS, there is a relatively large gap between DNS with full resolution and large eddy simulation (LES), which requires modeling of all fine, unresolved flow and chemical structures. In this sense, ODT may play an important intermediate role, by resolving structures in one dimension, with stochastically-modeled turbulent motions the compromise for cost efficiency. A comparison is made of the ODT model with a series of three parametric DNS simulations of turbulent, nonpremixed, planar ethylene jet flames. These three simulations occur at a constant Reynolds number, but varying Damkohler number (ratio of characteristic reaction rate to mixing rate), resulting in varying levels of flame extinction. Each simulation exhibits a high degree of turbulence-flame-chemistry interaction. The DNS simulations are presented, along with the ODT model. Comparisons between the models includes the spatial evolution of the jets, statistical spatial properties (mean and rms) of velocity, temperature, and composition profiles, and statistics of these properties conditioned on the nominal flame-normal coordinate (mixture fraction). Remakable agreement between the ODT and DNS models is observed with 3-4 orders of magnitude difference in computational cost.

See more of this Session: CFD and Mixing Effects in Chemical Reaction Engineering
See more of this Group/Topical: North American Mixing Forum

ANALYSIS OF THE MICRO-MIXING TIME IN FDF METHODS

Juan M. Mejia, Farid Chejne, Alejandro Molina, Universidad Nacional de Colombia, Medellín, Colombia
Amsini Sadiki, Institute of Energy and Power Plant Technology. Technischen Universität Darmstadt, Germany

Abstract

In this work the solution of the FDF transport equation is done via Monte Carlo method, for a confined turbulent liquid jet discharging a conserved scalar into a low-velocity water stream A sensibility analysis of the number of stochastic particles as well as the influence of the SGS mixing time constant were evaluated in this work. Simulation results are in good agreement with reported experimental data. However it was observed that the use of a single SGS mixing time constant is not appropriate for the configuration studied in this work, showing that this parameter should be modeled or dynamically calculated.

Introduction

Mixing is an important process in many engineering, biological and environmental applications. In reactive systems, chemical reactions can only take place after reactants are mixed at molecular level. In many industrial and biological applications taking place in liquid phase, the ratio of the kinematical viscosity to molecular diffusivity of the scalar is much greater than one, so the scalar field holds finer structures than the velocity field. The small dimension of the Batchelor scale makes accurate measurement of the instantaneous concentration in turbulent liquid flows scarce and only possible with rather sophisticated systems [1].

Direct Numerical Simulation (DNS) is the most direct way to get information about the fluid flow but the computational cost is prohibitive in most engineering applications. A more realistic alternative to DNS is the utilization of spatial filtering or temporal ensemble averages such as Large Eddy Simulation (LES) and Unsteady Reynolds-Averaged Navier Stokes equations (U-RANS). In both approaches, additional unknown terms appear in the final form of the balance equations and they must be modeled in order to close the numerical set of equations. Large Eddy Simulations (LES) is based on the concept of scale separation by using a low-pass filter, where the turbulent large scales of motion are solved while the subgrid components are modeled.

Mixing can be simulated by the numerical solution of the filtered continuity, momentum and scalar equations. Subgrid-scale (SGS) models must be supplied in order to take into consideration interactions between resolved and non-resolved parts of the turbulent spectrum. A second approach is to solve the joint probability function of the SGS scalars [2], named Filtered Density Function (FDF), in conjunction with the filtered momentum equation from LES. This approach has also a modeling requirement because the conditional diffusion term in the FDF transport equation is unknown. This important term accounts for transport in physical space and mixing in the composition space and it has to be modeled by mixing models.

A widely used mixing model is the Interaction by Exchange with the Mean (IEM) [3, 4], and will be used in this study. The IEM model has a deterministic origin, initially formulated for PDF methods in RANS context. The IEM states that the stochastic particle only interacts with itself, relaxing the solution in the composition space towards the mean value; the relaxation is done over a subgrid mixing time. Experimental results for gas flows show that the sub grid mixing time is not universal, with measured values between 0.6 and 3.1 in the RANS context [5]. In addition, molecular transport properties are not taken into account in these models. The non-universality of these constants can affect the predictive capability of LES [6]. In this work, a LES-FDF approach is carried out in order to reproduce the experimental results of a turbulent mixing process.

Formulation

The hybrid Eulerian/Lagrangian LES / FDF approach is based on the numerical solution of both the Eulerian filtered momentum equation via Large Eddy Simulation, and the joint scalar Filtered Density Function.

Large Eddy Simulation (LES)

LES can be viewed as an intermediate approach between DNS and RANS, because in the former all turbulent structures are calculated, while in the later they are modeled. The idea of the LES approach is to solve the large, non-universal, anisotropic turbulent scales and to model the short turbulent scales, nearly universal. Scale separation is done by applying a low-pass filter. Spatial fluctuations, lower than a defined filter cut-off length -Δ-, are smoothed or removed. The spatial filtering operation of a flow variable Q is defined by the convolution integral:

$$\langle Q(\boldsymbol{x},t)\rangle_l \equiv G\star Q = \int_D G(\boldsymbol{x}-\boldsymbol{x}')Q(\boldsymbol{x}',t)\,d\boldsymbol{x}' \qquad (1)$$

Where G is the filter kernel defined over the entire domain D. Box, Gaussian and Fourier filters are commonly used. The difference between a flow variable (Q) and its filtered (or resolved) value $(\langle Q(x)\rangle_l)$ is called the subgrid scale -SGS- (or unresolved) value (Q'): $Q=\langle Q\rangle_l + Q'$

The Favre filtering is defined by:

$$\langle Q\rangle_L \equiv \frac{1}{\langle\rho\rangle_l}\int_D \rho G(x'-x')Q(x)\,dx' = \frac{\langle\rho Q\rangle_l}{\langle\rho\rangle_l} \qquad (2)$$

Favre filtered mass, and momentum equations are:

$$\frac{\partial\langle\rho\rangle_l}{\partial t} = \frac{\partial}{\partial x_j}\left(\langle\rho\rangle_l\langle u_j\rangle_L\right) = 0 \qquad (3)$$

$$\frac{\partial\langle\rho\rangle_l\langle u_i\rangle_L}{\partial t} + \frac{\partial}{\partial x_j}\left(\langle\rho\rangle_l\langle u_i\rangle_L\langle u_j\rangle_L\right) = -\frac{\partial}{\partial x_j}\left[\langle\tau_{ij}\rangle_l + \tau_{ij}^{sgs}\right] - \frac{\partial\langle p\rangle_l}{\partial x_i} + \langle\rho\rangle_l F_i \qquad (4)$$

where ρ is the fluid density, u is the velocity, τ_{ij} is the stress tensor, τ_{ij}^{sgs} is the SGS stress tensor, p is the pressure and F is the large-range external forces. It can be seen that new terms appear in LES equations, as it occurs in RANS. These terms account for resolved and subgrid scales interactions. The SGS stresses are defined by:

$$\tau_{ij}^{sgs} = \left(\langle u_i u_j \rangle_L - \langle u_i \rangle_L \langle u_j \rangle_L\right)\langle \rho \rangle_l \tag{5}$$

A simple SGS model, widely used is the Smagorinsky model [7]:

$$\tau_{ij}^{r} = -2\nu_r \langle \rho \rangle_l \langle S_{ij} \rangle_L \quad , \quad \langle S_{ij} \rangle_L = \frac{1}{2}\left(\frac{\partial \langle u_i \rangle_L}{\partial x_j} + \frac{\partial \langle u_j \rangle_L}{\partial x_i}\right) \tag{6}$$

Where ν_r, is the eddy viscosity of residual motions. It is modeled in a similar way as the mixing length:

$$\nu_r = (C_s \Delta)^2 |S|_L \quad , \quad |S|_L \equiv \left(2\langle S_{ij} \rangle_L \langle S_{ij} \rangle_L\right)^{1/2} \tag{7}$$

C_s is the Smagorinsky coefficient that can be specified or dynamically calculated using the Germano procedure [8]

Filtered probability density function equation (FDF)

In FDF, a filtered probability density function (FDF) is used to quantify the probability to find a filtered variable of the flow (*Y*) in the range [*Y**-Δ*Y*/2 , *Y**+Δ*Y*/2] . The fundamental property of the FDF method is to account, in a probabilistic way, for the effects of SGS fluctuations.

The statistics of the reactive scalars, i.e. the FDF of the scalars, can be explicitly obtained from a transport equation. Pope [2] introduced the mathematical definition of the FDF transport equation, demonstrating that the chemical reaction term appears in a closed form and that; therefore modeling of this term is not required. Drozda et al. [9] present an overview of the state of progress in –FDF.

The mass filtered mass density function (FMDF) is defined as:

$$F_L(\psi; x, t) = \int_{-\infty}^{+\infty} G(x - x', t)\rho(x', t) f'(\psi; x', t)\mathrm{d}x' \tag{8}$$

where ψ is the sample space variable for each composition and f' is the fine-grained joint-PDF of compositions [10], defined as:

$$f'(\psi; x, t) \equiv \prod_{\alpha=1}^{n_\phi} \delta\left(\phi_\phi[x,t] - \psi_\phi\right) = \delta\left(\phi[x,t] - \psi\right) \tag{9}$$

where δ is the Dirac delta function. Based on the properties of Dirac delta functions and, after some manipulation, the FMDF equation is [6]:

$$\frac{\partial F_L}{\partial t} + \frac{\partial}{\partial x_j}\left[F_L \langle u_j | \psi \rangle_L\right] = -\frac{\partial}{\partial \psi_\alpha}\left[F_L \left\langle \frac{\mathrm{D}\phi_\alpha}{\mathrm{D}t} \middle| \psi \right\rangle_L\right] \tag{10}$$

The introduction of the scalar balance equation into equation (12) yields:

$$\frac{\partial F_L}{\partial t}+\frac{\partial}{\partial x_j}\left[F_L\left\langle u_j\middle|\psi\right\rangle_L\right]+\frac{\partial\left[F_L S_\alpha\right]}{\partial\psi_\alpha}=-\frac{\partial}{\partial\psi_\alpha}\left[F_L\left\langle-\frac{1}{\rho}\frac{\partial J_\alpha}{\partial x_j}\middle|\psi\right\rangle_L\right] \tag{11}$$

The zeroth, first and second order moments of the FMDF are:

$$\int F_L(\psi;x,t)\,d\psi=\langle\rho\rangle_l \tag{12}$$

$$\int \psi_\alpha F_L(\psi;x,t)\,d\psi=\langle\rho\phi_\alpha\rangle_l\equiv\langle\rho\rangle_l\langle\phi_\alpha\rangle_L \tag{13}$$

$$\int \psi_\alpha\psi_\beta F_L(\psi;x,t)\,d\psi=\langle\rho\rangle_l\left\langle\phi_\alpha\phi_\beta\right\rangle_L \tag{14}$$

Conditionally filtered terms of equations are unclosed. The conditional advection term can be modeled by the conventional gradient diffusion [6]:

$$\left(\left\langle u_j\middle|\psi\right\rangle_l-\left\langle u_j\right\rangle_L\right)F_L=-\langle\rho\rangle_l D_t\frac{\partial}{\partial x_j}\left(\frac{F_L}{\langle\rho\rangle_l}\right) \tag{15}$$

Where D_t is a SGS diffusion coefficient. The conditional diffusion term in Eq. (11), accounts for transport in physical space and mixing in the composition space. It has to be modeled by mixing models. The Interaction by Exchange with the Mean, IEM, model (or Linear Mean Square Estimation, LMSE) [3 4] has been usually used [see 6, 11-13]. Other less frequently used models is the modified Curls model (MCURL) [13]. These mixing models were developed in the RANS context. The IEM model can be described by Eq. (16) [6]:

$$\frac{\partial}{\partial\psi_\alpha}\left[\left\langle-\frac{1}{\rho}\frac{\partial J_\alpha}{\partial x_j}\middle|\psi\right\rangle_L F_L\right]\approx\frac{\partial}{\partial x_j}\left[\langle\rho\rangle_l\langle D\rangle_L\frac{\partial}{\partial x_j}\left(\frac{F_L}{\langle\rho\rangle_l}\right)\right]+ \\ \frac{\partial}{\partial\psi_\alpha}\left[\Omega_m\left(\psi_\alpha-\langle\phi_\alpha\rangle_L\right)F_L\right] \tag{16}$$

where Ω_m is the frequency of mixing within the subgrid and it can be modeled as $\Omega_m=C_\Omega\left(\langle D\rangle_L+D_T\right)/\Delta_L^2$, where C_Ω is the SGS mixing time constant. The solution of the FDF equation by conventional difference schemes is intractable due to the high dimensionality of the equation: the computational cost rises exponentially as scalars are increased [14]. To overcome this problem, stochastic methods have been used for the solution of the equation, because the computational cost rises linearly [13].

In this work we follow the Lagragian Monte Carlo procedure [5], which is described by Jaberi et al. [6] and that considers that there are a set of stochastic particles uniformly distributed in the computational domain. These particles evolve in physical and compositional space according to a set of stochastic differential equations SDEs. The moments of the scalar properties on an ensemble average should be equivalent to the moments of the FDF. The transport of the particles in the physical space is given by the following SDE:

$$\mathrm{d}X_i = D_i\left(\boldsymbol{X}(t),t\right)\mathrm{d}t + E\left(\boldsymbol{X}(t),t\right)\mathrm{dW}_i(t) \quad (17)$$

$$E = \sqrt{2\left(\langle D\rangle_L + D_T\right)/\langle\rho\rangle_l} \quad \text{(Drift coefficient)} \quad (18)$$

$$D_i = \langle u_i\rangle_L + \frac{1}{\langle\rho\rangle_l}\frac{\partial\left(\langle D\rangle_L + D_T\right)}{\partial x_i} \quad \text{(Diffusion coefficient)} \quad (19)$$

where $\boldsymbol{X}$ is the Lagrangian position, D_i is the drift coefficient, E is the diffusion coefficient and W_i is a Wiener process. When the IEM mixing model is used, the transport equation of the particles in the compositional space reads:

$$\frac{\mathrm{d}\phi_\alpha\left(\boldsymbol{X}(t),t\right)}{\mathrm{d}t}(t) = \frac{\mathrm{d}\phi_\alpha^+}{\mathrm{d}t} = -\Omega_m\left(\phi_\alpha^+ - \langle\phi_\alpha\rangle_L\right) \quad (20)$$

Numerical solution

The filtered momentum balance equations were solved using the finite-volume CFD code FASTEST-3D. The code uses geometry-flexible, block-structured, boundary-fitted grids and cell-centered variable storage. Spatial discretization considered upwind schemes for first order terms second order central schemes for second order terms. A SIMPLE type pressure correction method was used for pressure-velocity coupling and an iterative strongly implicit procedure (SIP) solver for solving the resulting set of linear equations. Time integration was carried out using the second order implicit Crank-Nicholson method. The code allows for parallelization based on domain decomposition using MPI message passing library. For more details, please refer to Durst and Schaefer [15].

In this study, a FDF solver was developed for the original LES code. The domain was decomposed in cells, and the ensemble average of *NPC* particles per cell represents the FDF equation. These cells, or finite volume, are used for constructing the statistics of the FDF. The LES solver provides the filtered velocity and SGS viscosity to the FDF solver and then Equations (17) and (20) are numerically integrated using the Euler method. Multiple time integrations can be performed in one LES time step.

In order to integrate Eq. (17), drift and diffusion coefficients Eq. (18-19) are interpolated from the LES solver onto particle position, by using tri-linear interpolation. The filtered scalar value $\langle\phi_\alpha\rangle_L$ in equation (20) is also interpolated from the FDF mesh for the integration of Eq. (20). Any filtered value $\langle Q\rangle_L$ is constructed on the basis of the assemble average on a FDF cell:

$$\langle Q\rangle_L \approx \frac{\sum_{i=1}^{NPC} w^i Q^i}{\sum_{i=1}^{NPC} w^i} \quad (21)$$

where w^i is the weight of the particle i. The weight of each particle is established according to the instantaneous local mass flow rate at the inflow boundaries. Considering the existence of velocity gradients and that the size of the FDF cell may not be constant, the number of particles in each cvaries.

In order to overcome this problem a clustering – splitting algorithm was implemented in order to keep NPC around a desired value (NPC ± 15).

Results and Analysis

We studied the experimental configuration from Antoine et al., [16] , in which a round jet (D=1 mm) discharged water (U_0=10 m s-1) into a low velocity co-flowing water stream (U_0 = 0.5 m s^{-1}). The jet is doped with rhodamine B (concentration: 5 × 10^{-6} mol l^{-1}) into a square section of 63 mm side and 1.5 m length. Combined Laser-Induced Fluorescence and 2D Laser Doppler Velocimetry techniques allowed measurement of velocity and molecular concentration.

LES were conducted on a computational domain of 100D × 63D × 63D (x × y × z). From a previous study [17] a good compromise between computational time and accuracy was obtained using a computational mesh containing 312 × 62 × 62 finite volumes in (x × y × z) directions, respectively. The cell size in the axial direction (x) increases as the distance from the nozzle becomes longer. For the lateral directions (y and z), the mesh is refined at the jet centerline and at the solid boundary, but has a lower resolution in between. The time step used in the simulations was 4.9 × 10^{-6} s.

The FDF solver runs on a frustum cone, where the bottom and top radius are 4D and 17D, respectively. The height of the cone is 80 D. The FDF mesh has a cylindrical coordinate system, refined on areas near the center. This coordinate system was chosen because more details can be represented in areas where strong gradients take place such as he jet axis centerline and surroundings.

Time-averaged statistics were collected over two flow-times in order to get statistical convergence. Initially, three number of particles per cell (25, 35 and 50) were evaluated in order to choose one for further simulations. The FDF mesh consist on a 300 × 20 × 15 nodes in axial (x), radial (r) and angular (θ) directions respectively and C_Ω= 8, as reported for gas flows. Figure 1 shows the streamwise concentration distribution for 25, 35 and 50 particles per cell.

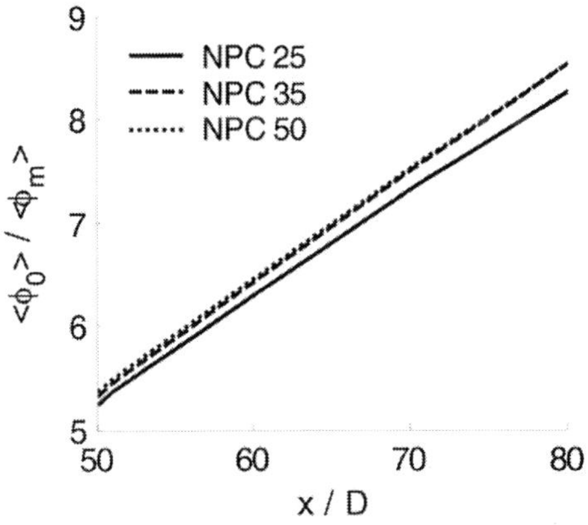

Figure 1. Simulation of the streamwise concentration distribution along the jet centerline, with NPC as parameter

In Figure 1 it is clear that the FDF simulations using 35 and 50 particles per cell yield similar results, therefore NPC = 35 was used for all the simulations presented below. Considering that the radial distributions given by the simulations (not shown here) for different axial positions obey the self

– similarity concept, simulations results are presented only for one axial location to improve readability of the plots. Contrary, the experimental data are presented for different axial locations (x/D). Figures 2 and 3 illustrate the effect of the FDF mesh size on the simulation results. In these figures η is the self-similarity coordinate, defined as η= r / x, while the label on each figure denotes the number of nodes in $x \times r \times \theta$.

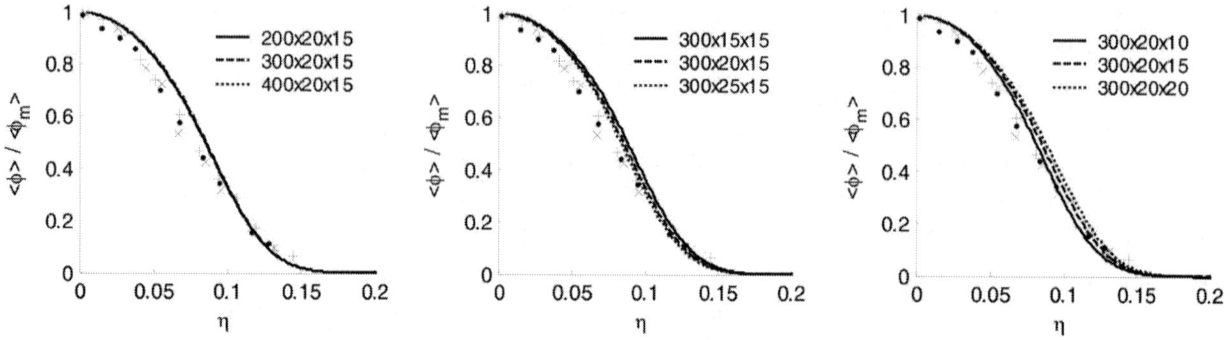

Figure 2. Radial mean concentration distribution across the jet for different nodes in axial (left) radial (center) and angular (right) nodes. FDF simulation results (lines). Experimental data [16] (symbols): + x/D = 70; • x/D = 80; × x/D = 90.

In Figure 2 the agreement of the simulations with experimental observations is good, and predictions are very similar for different mesh sizes.. The concentration is over predicted in the range of η [0.02 – 0.1]. Figure 2 reveals that the number of nodes in axial direction (for the range evaluated in this work) has no effect on the radial mean concentration distribution. It is also observed in Figure 2 that increasing the number of nodes in radial direction slightly enhances the prediction. On the contrary, the angular component is more sensitive to simulations.

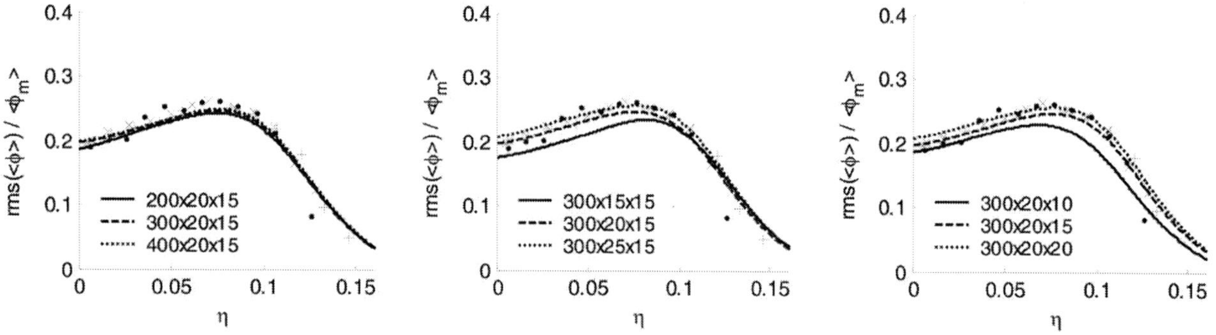

Figure 3. Radial distribution of the RMS of the concentration across the jet for different nodes in axial (left) radial (center) and angular (right) nodes. Simulation results (lines). Experimental data [16] (symbols): + x/D = 70; • x/D = 80; × x/D = 90.

The fluctuating component of the concentration is well captured by the FDF simulations (Figure 3). The model follow the trends in the number of nodes in axial direction, previously observed for the radial mean concentration distribution. The radial component has a significant impact on the simulation results. Increasing the number of nodes the predictions are improved. However, increasing from 20 to 25 radial nodes the predictions is to some extent improved, rising the computational cost. The later is also observed for the angular component.

From Figures 2 and 3, in can be said that the predictions of the FDF method are very good, even more than LES simulations using advanced SGS scalar flux models, such as the anisotropy model [18], shown in Figure 4.

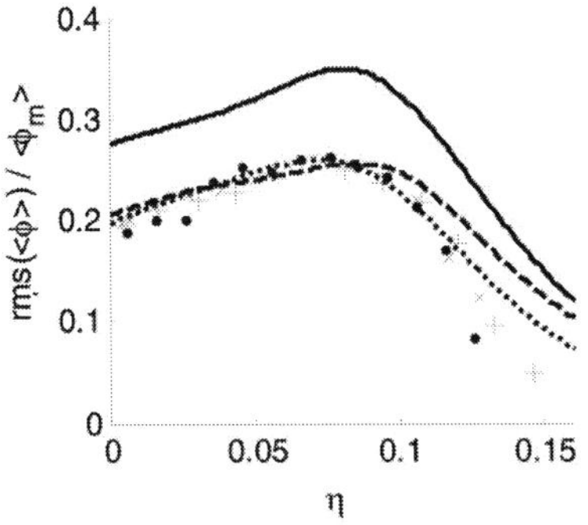

Figure 4. Radial distribution of the RMS of the concentration across the jet. LES simulation results [17] (lines): Solid: Eddy diffusivity model [19]; dashed: Eddy diffusivity model with dynamic Sc_T [20]; dotted: Anisotropy model [18]. Experimental data [16] (symbols): + x/D = 70; ● x/D = 80; × x/D = 90.

So far a single value for the computation of the mixing time in Eq. (16) has been used ($C_\Omega = 8$). In Figure 5 simulations were done on a 300 × 20 × 15 FDF mesh using four values for the SGS mixing time constant.

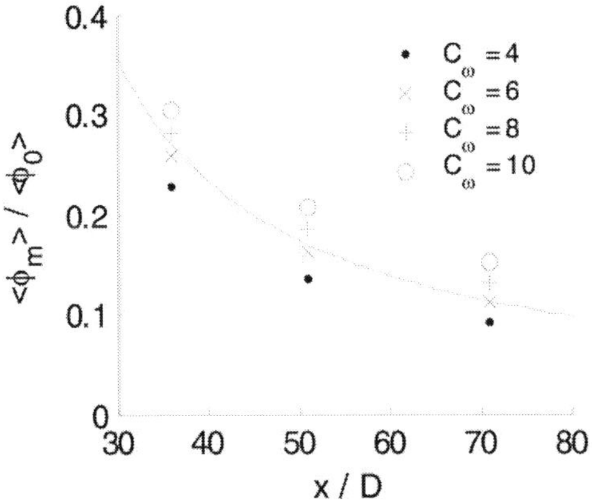

Figure 5. Streamwise mean concentration distribution. Simulation (symbols). Theoretical profile [16] (line):

Figure 5 shows that the FDF simulation results trend the theoretical distribution, when the SGS mixing time constant is in the range [6-8]. Another important result is that Figure 5 reveals that the use of a single SGS mixing time constant is not suited for the flow studied in this work. The non-universality of this constants can affect the predictive capability of FDF.

Summary and concluding remarks

The simulation of a mixing process taking place in liquid phase was done using a hybrid LES – FDF approach. A solver, based on the Lagnagian Monte Carlo method was developed in order to solve the FDF transport equation. The FDF solver is coupled to the Eulerian FASTEST 3D for the numerical solution of the filtered momentum equation. It was observed that 35 stochastic particles per FDF cell give similar results to that using 50, at a lower computational time. By changing the number of FDF nodes in axial, radial and angular directions seven meshes were tested and their results were compared to experimental data of a turbulent liquid jet (Re=10000) doped with a conserved scalar. The simulations results showed good agreement with experimental observations. The simulation results did not appreciably changed for the range of nodes in axial direction used un this work. On the contrary, results were more sensitive to changes in radial and angular directions. An important observation given by the present study is that the FDF predictive capabilities are dependent on the SGS mixing time constant, which a single constant value is not suited for the present configuration.

Acknowledgments

Authors thank the DAAD-COLCIENCIAS under collaboration PROCOL 2007 for a traveling grant. The first author acknowledges especially the Institute of Energy and Power Plant Technology (EKT) at the Technische Universität Darmstadt (Germany) where a part of this work was conducted. Authors thank also the Faculty of Mines (U.N.C.) for traveling support.

References

1. Michioka T, Komori S. (2004). "Large-Eddy Simulation of a Turbulent Reacting Liquid Flow". *AIChE Journal*, 50 (11), pp. 2705-2720.
2. Pope, S.B. (1990). "Computations of turbulent combustion: Progress and challenges". *Proceeding of the Combustion Institute*, 23 pp. 591–612.
3. Dopazo, C. and O'Brien, E. E. (1976). "Statistical treatment of non-isothermal chemical reactions in turbulence". *Combustion Science and Technology*, 13, pp 99-112.
4. O'Brien, E. E. (1980). "The probability density function (PDF) approach to reacting turbulent flows". In *Turbulent Reacting Flows* (ed. P. A. Libby & F. A. Williams), chap. 5, pp. 185-218. Springer.
5. Pope S.B. (1985). "PDF methods for turbulent reactive flow". *Progress in Energy and Combustion Science,* 11, pp. 119-192.
6. Jaberi S. A., Coluci P.J., James S., Givi P., Pope S.B. (1999). "Filtered mass density for large-eddy simulation of turbulent reacting flows". *Journal of Fluid Mechanics*, 40, pp. 85-121.
7. Smagorinsky J.S. (1963) "General circulation experiments with the primitive equations I. The basic experiment". *Monthly Weather Review*. 91 (3), pp 99-164.
8. Germano M, Piomelli U, Moin P, Cabot W.H. (1991). "A dynamic subgrid-scale eddy viscosity model". *Physics of Fluids*. 3(7), pp. 1760-1765.
9. Drozda, T.G., Sheikhi, M.R.H., Madnia, C.K., Givi, P., (2007). "Developments in formulation and application of the filtered density function". *Flow Turbulence and Combustion*, 78, pp. 35-67.
10. Pope, S.B. (2000). *Turbulent flows*. Cambridge University Press, Cambridge.

11. van Vliet E., Derksen J. J., van den Akker H. E. A. (2005). "Turbulent Mixing in a Tubular Reactor: Assessment of an FDF/LES Approach". *AIChE Journal*, 51 (3), pp. 725-739.
12. Mustata R., Valiño L., Jiménez C., Jones W.P., Bondi S. (2006). "A probability density function Eulerian Monte Carlo field method for large eddy simulations: Application to a turbulent piloted methane/air diffusion flame (Sandia D) ". *Combustion and Flame*, 145, pp. 88–104.
13. Olbricht, C., Hahn, F., Sadiki, A., Janicka, J. (2007). "Analysis of subgrid scale mixing using a hybrid LES-Monte-Carlo PDF method". *International Journal of Heat and Fluid Flow*, 28, pp. 1215-1226
14. Jones W.P., Navarro-Martinez S. (2007). "Large eddy simulation of autoignition with a subgrid probability density function method. *Combustion and Flame,* 150, pp. 170–187.
15. Durst F, and Schäfer M. (1996). "A parallel blockstructured multigrid method for the prediction of incompressible flow". *International Journal for Numerical Methods in Fluids*, 22, pp. 549–565.
16. Antoine Y, Lemoine F, Lebouché M. (2001). "Turbulent transport of a passive scalar in a round jet discharging into a co-flowing stream". *European Journal of Mechanics: B/Fluids*, 20, pp 275-301.
17. Mejía J.M., Sadiki A., Molina A., Chejne F., Parra J.C. Pantangi P. (2010). "Evaluation of SGS scalar flux models in a turbulent liquid flow". AIChE annual meeting, Salt Lake City, Utah.
18. Sadiki A, Jakirlic S, Hanjalic K. (2003). "Towards a thermodynamically consistent, anisotropy-resolving turbulence model for conjugate flow, heat and mass transfer". In *Turbulence, Heat and Mass Transfer* (Ed. Hanjalic K., Nagano Y. and Trummers M..J.), 4. Begell House, Inc., pp. 545-551.
19. Eidson T. (1985). "Numerical simulation of the turbulent Rayleigh-Bénard problem using subgrid modeling. *Journal of Fluid Mechanics*, 158, pp.245-268.
20. Lilly D. (1992). "A proposed modification of the Germano subgrid-scale closure method". *Physics of Fluids*, 4 (3), pp. 633-635.

(522d) Approach towards a Quantitative Description of Microstructured Cyclone Type Mixers

Andreas Kölbl[1], Manfred Kraut[2], Achim Wenka[2] and Roland Dittmeyer[2], (1)Institute for Micro Process Engineering (IMVT), Karlsruhe Institute of Technology, 76344 Eggenstein-Leopoldshafen, Germany, (2)Imvt, Karlsruhe Institute of Technology, 76344 Eggenstein-Leopoldshafen, Germany

Approach Towards a Quantitative Description of Microstructured Cyclone Type Mixers

Authors

A. Kölbl*, M. Kraut, A. Wenka, and R. Dittmeyer:

Institute for Micro Process Engineering, Karlsruhe Institute of Technology, 76344 Eggenstein-Leopoldshafen, Germany.

*Telephone: 07247/82-6653, Fax 07247/82-3186, Email: Andreas.Koelbl@kit.edu

Mixing of two liquids in miniaturized Cyclone Type Mixers (figure 1) is achieved by feeding them tangentially into a mixing chamber.

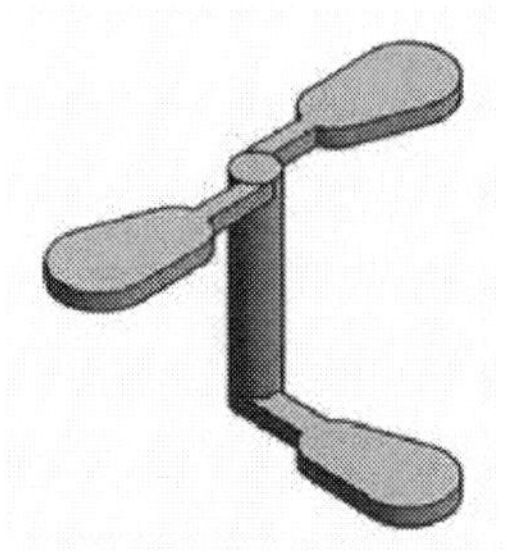

Figure 1: Principal sketch of Cyclone Type Mixers.

Thus a swirl motion is created, which enhances mixing. We recently published a parameter study on the qualitative influence of various structural parameters on mixing in such devices [Kölbl2010] based on computational fluid dynamics (CFD) and experiments using the Iodide Iodate Reaction Method. It was found that the diameter of the cyclone mixing chamber is more important for mixing than the mixing chamber length. Furthermore, the results from numerical simulations indicate a swirl motion in the mixing chamber as well as the occurrence of cavitation and zones of complete recirculation (dead zones).

In this contribution we develop a quantitative description of mixing in Cyclone Type Mixers. The influence of the mass flow throughput on the 'mixing quality' is discussed as well as the influence of the viscosity on hydrodynamics, pressure drop, and mixing.

Influence of the mass flow rate on mixing

The results from CFD calculations (Fluent 12.1) indicate that the distribution of a passive inert tracer is almost independent on the mass flow rate in the mixing device. Figure 2 shows the normalized distribution of a passive inert tracer at two different mass flow rates: (a) 2x1.5 kg/h and (b) 2x3 kg/h.

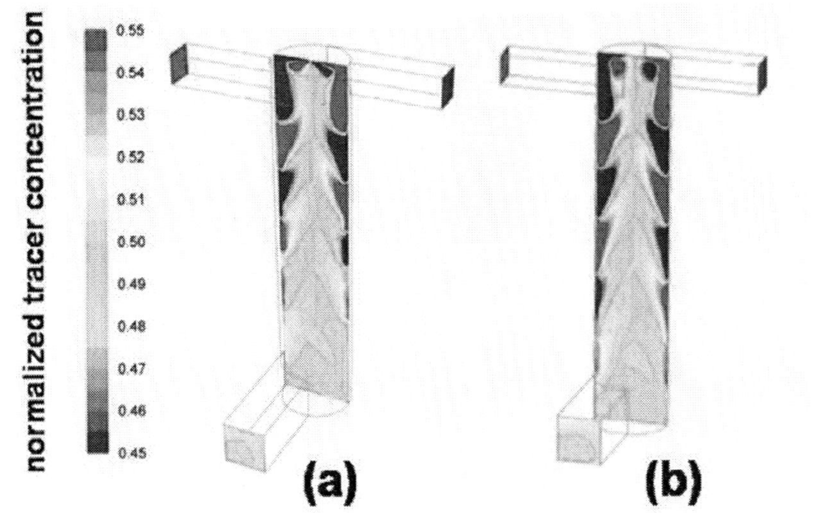

Figure 2: Distribution of a passive inert tracer at two different mass flow rates: (a) 2x1.5 kg/h and (b) 2x3 kg/h.

The findings suggest that the mean mixing time, defined here after Lindenberg et al. [Lindenberg2008] as the residence time along the stream lines until complete mixing is reached can be directly adjusted by the mass flow rate. An experimental verification of this scaling assumption is provided by applying the Iodide Iodate Reaction Method. Assuming that the triiodide concentrations in the resulting mixtures are directly proportional to the mixing times [Falk2010], the triiodide concentrations should be proportional to $(dm/dt)^{-1}$. This assumption is verified through the observation that the triiodide concentrations in the resulting mixtures can be described by a simple empirical correlation (equation 1) for various devices and concentration sets.

(1) c(triiodide) = A/(dm/dt)

Thus the triodide concentrations only depend on a single parameter A, which is a function of the reactant concentrations, the viscosity, and geometrical details of the examined mixing devices.

Influence of the viscosity on hydrodynamics, pressure drop, and mixing

Results from numerical simulations further indicate that the pressure drop in the mixing chamber, which is decisive for the mixing results is independent of the viscosity. It should be kept in mind that the pressure drop in the mixing chamber which includes feed and exit channels and fluid bends accounts only for about 50 % of the whole device [Kölbl2010]. However, the number of times the fluids rotates in the mixing chamber is reduced at higher viscosities. Thus the contact area between the fluids is diminished considerably. Figure 3 shows path lines for a fluid with a viscosity of (a) 0.5 mPas and (b) 10 mPas in the same device. The entry velocity is 5 m/s in both cases.

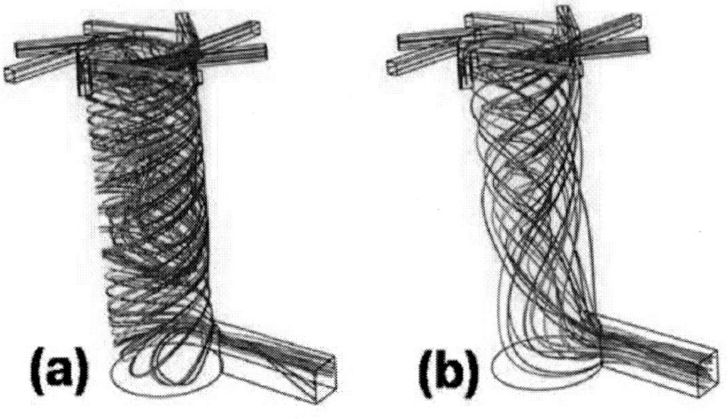

Figure 3: Pathlines for fluids of different viscosities: (a) 0.5 mPas and (b) 10 mPas.

Due to the reduction of fluid turns in the mixing chamber, reduced mixing qualities are expected for fluids at elevated viscosities. This expectation is verified by applying the Iodide Iodate Reaction Method for viscosities of 1 mPas (aqueous solution) and 2 mPas (aqueous solution + glycerol). For Cyclone Type Mixers the ratio of the optical densities of the resulting solutions obtained at 2 mPas and 1 mPas is independent of the flow rates applied. This is another experimental finding that supports the sacling assumption from CFD calculations. However the influence of the elevated viscosity is rather small compared to microstructured V-type mixers [Kölbl2009].

[Kölbl2010]

A. Kölbl, M. Kraut, and A. Wenka:

"Experimental Parameter Study on Cyclone Type Mixers"

11th Conference on Micro Reaction Technology (IMRET 11), Kyoto, Japan, March 8.-10. 2010.

[Lindenberg2008]

C. Lindenberg, J. Schöll, L. Vicum, M. Mazotti, and J. Brozio:

„Experimental characterization and multi-scale modeling of mixing in static mixers"

Chemical Engineering Science 63(2008)4135-4149.

[Falk2010]

L. Falk and J.-M. Commenge:

„Performance comparison of micromixers"

Chemical Engineering Science 65 (2010) 405-411.

[Kölbl2009]

A. Kölbl, M. Kraut und K. Schubert:

„Mischexperimente mit einem Mikromischer bei verschiedenen Viskositäten"

Chemie Ingenieur Technik 81(2009)1058-1059.

See more of this Session: CFD and Mixing Effects in Chemical Reaction Engineering
See more of this Group/Topical: North American Mixing Forum

EVALUATION OF SGS SCALAR FLUX MODELS IN A TURBULENT LIQUID FLOW

Juan M. Mejia, Alejandro Molina, Farid Chejne, Camilo Parra, Universidad Nacional de Colombia, Medellín, Colombia
Amsini Sadiki, Pradeep Pantangi, Institute of Energy and Power Plant Technology. Technischen Universität Darmstadt, Germany

Abstract

It is well known that turbulent mixing plays an important role in various engineering and environmental applications. When numerical simulations of industrial processes are carried out, key requirements are the computational efficiency, algorithm robustness and accurate representation of the process. In this work, a Large Eddy Simulation (LES) of a mixing process is done where three different approaches for the subgrid-scale (SGS) scalar flux models are used for closing the filtered scalar balance equation: eddy diffusivity model with both constant and dynamically-calculated turbulent Schmidt number (Sc_T) and the dynamic anisotropy model. The later model was derived in compatibility with the second law of thermodynamics and combines in its simplest form the eddy diffusivity model with an additional term that couples the (deviatoric) SGS stress tensor and the gradient of the filtered scalar field. This term is a measure of how the local mixing, dependent on the molecular Schmidt (Sc) number through the SGS time scale, is influenced by the non-resolved flow structures. Simulation results where validated with experimental data of a turbulent liquid round jet. A better agreement was achieved when the anisotropy model was used..

Introduction

Models that correctly represent the mixing processes in turbulent flows are important for the design of various engineering and environmental applications where turbulence plays an important role. While the current understanding of turbulence allows a reasonable description of the flow field [1], a detailed representation of mixing processes remains a challenge. However, a correct prediction of mixing is of paramount importance, particularly in the case of reactive systems for which chemical reactions can only take place after reactants are mixed at molecular level.

Numerous studies have dealt with transport and mixing of passive scalars in systems where the Schmidt number is close to one, as it is the case for gas flows in which the smallest scalar scales are of the same order of magnitude as the Kolmogorov scale. However, in many industrial and biological applications that take place in the liquid phase the ratio of the kinematical viscosity to molecular diffusivity of the concentration is much greater than one, so the scalar field holds finer structures than the velocity field and the Batchelor scale is much smaller than that in gas turbulence. The small dimension of the Batchelor scale makes accurate measurement of the instantaneous concentration in turbulent liquid flows scarce and only possible with rather sophisticated systems [2]. Important information about velocity and scalar fields can be obtained from numerical simulation, that is needed for design and process improvement.

Direct Numerical Simulation (DNS) gives the most detailed information about the flow as it resolves all the scalar length- and time-scales by setting up the spatial resolution according to the smallest scalar/flow scale. However, the computational cost rises as Re^3 and becomes very expensive, and even prohibitive, for systems of practical importance. A more realistic alternative to DNS is spatial filtering or temporal ensemble averages such as Large Eddy Simulation (LES) and Reynolds Averaged Navier-Stokes equations (RANS), respectively. In RANS the largest scales are solved and the turbulent-spectrum information is provided by turbulence models. Although RANS provides a reasonable computational cost -precision ratio, its predictions about the concentration fluctuations are not always well-captured when liquid flows are studied [2]; in the case of stirred tank reactors (STR) the turbulent kinetic energy is underpredicted in the impeller region [3].

LES can be viewed as an intermediate approach between DNS and RANS, because in the former all turbulent structures are calculated, while in the later they are modeled. The idea of LES is to solve the large, non-universal, anisotropic turbulent scales and to model the short turbulent scales, nearly universal. Scale separation is done by applying a low-pass filter. LES has been successfully applied to turbulent flows, and in particular to complex geometries in liquid phase, such as STR (see: [2-6]). Because LES equations are unclosed, SGS models must be supplied in order to specify the SGS stresses, SGS scalar fluxes and filtered reaction terms. For the scalar field different SGS scalar fluxes have been proposed in literature, including the isotropic-eddy diffusivity model [7] which is very popular due to its high simplicity/accuracy ratio for gas flows.

According to experimental data, the assumption of isotropy at inertial and dissipation scales is no longer valid for structured functions and derivative skewness of the scalar field in presence of a mean scale gradient [8]. Different models have been proposed in order to take into account the anisotropic behavior of micro-mixing, mainly focused on gas-phase, (see [9,10]), but their application for fluids with high Sc numbers has not been successful. In the case of STR systems, turbulent anisotropy plays an important role in the impeller stream, boundary layers and at the separation points, so the use of isotropic-based turbulence models should be treated with care [3].

In the present study different SGS scalar flux models, including the dynamic anisotropy model [11-13], are evaluated for a high Sc turbulent flow and compared to experimental data from Antoine et al., [14] of a turbulent round jet discharging a diluted scalar into a low velocity co-flowing liquid water .

Formulation

The isotermic flow field can be described by mass, momentum and species transport equations. In LES, the separation of large from small scale structures is done by applying filtering operations to the transport equations. For an incompressible, non-reacting isothermal flow the filtered continuity (1), momentum (2) and species (3) equations can be expressed as follows:

$$\frac{\partial \overline{U}_j}{\partial x_j} = 0 \tag{1}$$

$$\frac{\partial \overline{U}_i}{\partial t} + \frac{\partial}{\partial x_j}\left(\overline{U}_j \overline{U}_i\right) = \frac{1}{\rho}\frac{\partial}{\partial x_j}\left(2\mu \overline{S}_{ij} - \frac{2}{3}\mu \frac{\partial \overline{U}_k}{\partial x_k}\delta_{ij} + \tau_{ij}^{sgs}\right) - \frac{1}{\rho}\frac{\partial \overline{p}}{\partial x_i} \tag{2}$$

$$\rho \frac{\partial \overline{\phi}}{\partial t} + \frac{\partial}{\partial x_j}\left(\rho \overline{\phi}\,\overline{U}_j\right) = \frac{\partial}{\partial x_j}\left(\rho D_\phi \frac{\partial \overline{\phi}}{\partial x_j} - \rho J_j^{sgs}\right) \tag{3}$$

Where the over-bar denotes a filtered value, U_i is the velocity components at x_i direction, ρ is the fluid density, p is the pressure, ϕ is the concentration, μ is the fluid viscosity and D_ϕ is the molecular diffusion coefficient. The filtered, large-scale strain-rate tensor, $\overline{S}_{ij}$, is given by:

$$\overline{S}_{ij} = \frac{1}{2}\left(\frac{\partial \overline{U}_i}{\partial x_j} + \frac{\partial \overline{U}_j}{\partial x_i}\right) \tag{4}$$

The SGS stresses, $\tau_{ij}^{sgs} = \left(\overline{U_i U_j} - \overline{U}_i \overline{U}_j\right)$, in Eq. (2) are closed by using the Smagorinsky model [15]:

$$\tau_{ij}^{sgs} = -2\left(C_s \Delta\right)^2 |S| \rho \overline{S}_{ij} \tag{5}$$

Where Δ is the filter width, C_s is the Smagorinsky coefficient which is dynamically calculated by using the Germano procedure [16] and $|S| = \left(2\overline{S}_{ij}\overline{S}_{ij}\right)^{1/2}$.

SGS scalar flux models

The interaction between resolved and non-resolved scalar structures is accounted for by the unknown SGS scalar flux term, J_j^{sgs}, in Eq (3), which also must be provided via SGS scalar flux models. Macro, meso and micro-mixing happen simultaneously as the mixing process is taking place. The spatial and temporal transport of large–scale structures describes the macro-mixing and they can be solved in LES. Turbulent fluctuations in the energy-containing range of the velocity spectrum are, mostly, responsible for meso-mixing, while micro-mixing is driven by viscous–convective deformations of fluid elements and molecular diffusion. In the non-resolved inertial range, this can be expressed by gradient correlations. Mixture structures at small scales can be characterized by expressing the variance of the local concentration distribution around its average value [11-12]. This scalar variance can be understood as a measure of the departure of the scalar field from small–scale homogeneity, and its production depends on the known scalar flux.

Three different closure models for the SGS scalar flux, $J_j^{sgs} = \overline{U_j \phi} - \overline{U_j \phi}$, are traditionally used: the eddy diffusivity model where the turbulent Schmidt number (Sc_T) is assumed constant, the eddy diffusivity model using the dynamic procedure [17] for the estimation of the Sc_T and the dynamic anisotropy model. The eddy diffusivity model states that the SGS fluxes are proportional to the resolved scalar gradient:

$$J_j^{sgs} = -\nu_T / Sc_T \frac{\partial \overline{\phi}}{\partial x_j} \tag{6}$$

The model has been often used due to its simplicity, and it has shown good results for gas flows. Contrary to the eddy diffusivity and dynamic mosles, the anisotropic model [11-13] accounts for the, non-linear contributions of the SGS to the turbulent scalar flux. If it is assumed that the turbulent flow field (mean velocities and turbulence characteristics) is available from LES using the dynamic–based Smargorinsky SGS model [16], the anisotropy model used here is the simplest explicit anisotropic-resolving algebraic form, and consists of a cubic formulation in terms of the scalar gradients. This model is thermodynamically consistent as it agrees with the irreversibility requirement of the second law of thermodynamics [12,18]. It combines the linear eddy diffusivity model with an additional term coupling the (deviatoric) SGS stress tensor and the gradient of the filleted scalar field. The additional term is a measure of how the local mixing, dependent on the molecular Schmidt (Sc) number through the SGS time scale, is influenced by the non-resolved flow structures.. The anisotropy model is expressed as:

$$J_j^{sgs} = -\nu_T / Sc_T \frac{\partial \overline{\phi}}{\partial x_j} + D_{an} \Delta^2 \overline{S}_{ij} \frac{\partial \overline{\phi}}{\partial x_i} \tag{7}$$

Where D_{an} is the anisotropy model coefficient. In this work D_{an} is calculated by the dynamic procedure as well as the Sc_T. Preliminary evaluation of this procedure was reported in [12] for chemical liquid flows and gases in [,18], however in this paper we thoroughly evaluate the model performance for a different configuration.

Simulated Configuration and Numerical details

Up to date, turbulence and mixing characteristics in jet flows have been widely investigated as they have a simple geometry and are of practical importance in many industrial applications (e.g. fuel injectors, propulsion, chemical reactors). A jet results from discharging a fluid trough a circular orifice into a space. A free shear layer is created between the jet and the surroundings producing vortex sheets due to Kelvin-Helmholtz instability. We studied the experimental configuration from Antoine et al., [14] , see Figure 1, in which a round jet (D=1 mm) discharges water (U_0=10 m s^{-1}) into a low velocity co-flowing water stream (Uc = 0.5 m s^{-1}). The jet is doped with rhodamine B (concentration: 5×10^{-6} mol l^{-1}) into a square section of 63 mm side and 1.5 m length. Combined Laser-Induced Fluorescence and 2D Laser Doppler Velocimetry techniques allowed measurement of velocity and molecular concentration.

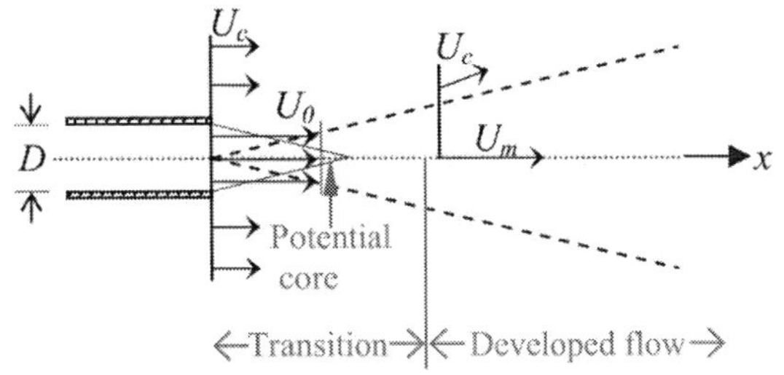

Figure 1. Scheme of the axis-symmetric jet [14].

The filtered momentum and scalar balance equations were solved using the finite-volume CFD code FASTEST-3D. The code uses geometry-flexible, block-structured, boundary-fitted grids and cell-centered variable storage. Spatial discretization considered upwind schemes for first order terms second order central schemes for second order terms. A SIMPLE type pressure correction method was used for pressure-velocity coupling and an iterative SIP solver for solving the resulting set of linear equations. Time integration was carried out using the second order implicit Crank-Nicholson method. The code allows for parallelization based on domain decomposition using MPI message passing library. For more details, please refer to Durst and Schaefer [18].

LES were conducted on a computational domain of 100D × 63D × 63D (x × y × z). Three meshes were analyzed in the first part of the study in order to evaluate the effect of mesh size on the velocity field. Table 1 shows the details of the three meshes. The cell size in the axial direction (x) increases as the distance from the nozzle becomes longer (see Figure 2). For the lateral directions (y and z), the mesh is refined at the jet centerline and at the solid boundary, but has a lower resolution in between. The time step used in the simulations was 4.9×10^{-6} s.

Table 1. Details of the computational meshes.

Label	0.8M	1.2M	1.6M
N_x	275	312	346
N_y, N_z	54	62	68
$N_{F.V.}$	801900	1199328	1599904
Δx,min (mm)	0.12	0.11	0.096
Δy,min (mm) Δz,min	0.038	0.032	0.031

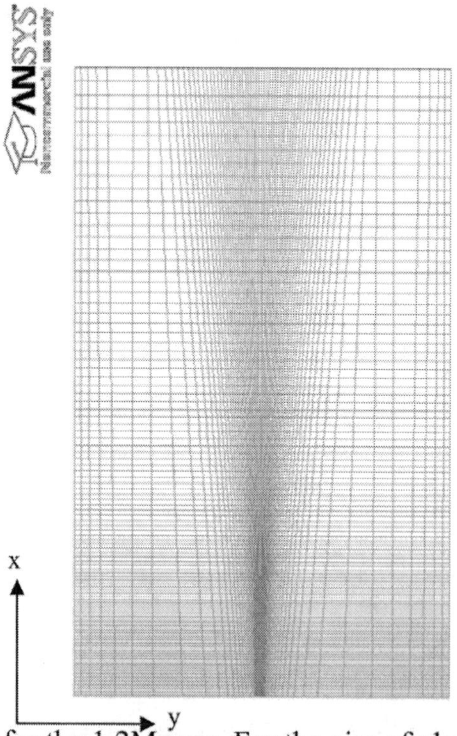

Figure 2. Mesh details at z=0 for the 1.2M case. For the aim of clearness, the finite volumes in the figure were scaled by a factor of 3.

At the nozzle exit, experimental data from Mi et al. [20] for radial mean and variance velocity profiles were used as inflow conditions. Random Gaussian fluctuations were added to the mean profiles. A zero-gradient condition was imposed at the downstream axial boundary, and non-flow (walls) conditions at the lateral boundaries.

Results and discussion

Mesh size sensitivity

Experimental observations of jet flows indicate that the spreading of the jet is linear, and that the inverse of the centerline streamwise velocity should be a straight line:

$$\frac{\langle U_0 \rangle}{\langle U_m \rangle - U_c} = \frac{1}{K}\left(\frac{x - x_0}{D}\right) \tag{8}$$

where K is the jet decay factor and x_0 is the virtual origin. Figure 3 reports the streamwise velocity decay distribution as a function of the distance to the orifice, obtained from the LES of 0.8M, 1.2M and 1.6M meshes. Figure 3 also presents experimental data from Antoine et al., [14].

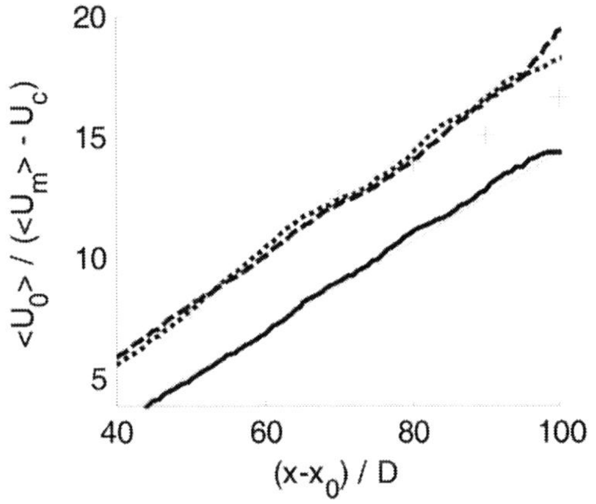

Figure 3. Streamwise mean velocity distribution along the jet centerline. Experimental data [14] (symbol). Simulation results (lines): Solid 0.8M mesh; dashed: 1.2M mesh; dotted 1.6M mesh. Gray lines denote linear regression.

Figure 3 shows that while the 0.8M mesh does a poor job in predicting the experiment data, the 1.2M and 1.6M cases show good agreement between experiments and simulation.In fact, the computed virtual origins, x_0, and jet decay factors, K, were similar with both meshes: 1.2M (x_0 = 6.64 D, K = 4.35) and 1.6M: (x_0 = 5.13D, K = 4.38). Comparison with experimental observations (K=6.83, x0= 4.9D) suggests that the distance where the potential core of the jet is influenced by shear layers is shorter. The virtual origins were over predicted in 35.5% and 4.6% for 1.2M and 1.6M meshes, respectively, suggesting that the mesh resolution may be a key aspect for this variable under the present configuration. The jet decay factor was under predicted by 36.3% and 35.9% for both cases. The departure from experiments, although high, is typical for experimental and numerical simulation of these systems and, in any case, is considerable acceptable for the evaluation of the effect of the SGS model in the final LES performance and should capture the qualitative behavior of the system.

For $(x-x_0/D) > 50$ 1.2M and 1.6M the simulation results do not follow a straight line but the tendency does (the ratio of the sum of squares of the regression -R^2- is 0.997 and 0.998, respectively). The latter advice that more computational time is needed in order to get statistical convergence, but it could not be done due to computational limitations.

Both the mean streamwise velocity (Figure 4) and the velocity fluctuations (Figure 5) obtained in the 1.2M and 1.6M are very close and their behavior agrees with experimental observations. For all meshes, the turbulent kinetic energy contained in the streamwise direction is redistributed towards the radial component, however, in Figure 5 the 0.8M case does not correctly capture the turbulent kinetic energy in the axial direction in the centerline (Figure 5 a).

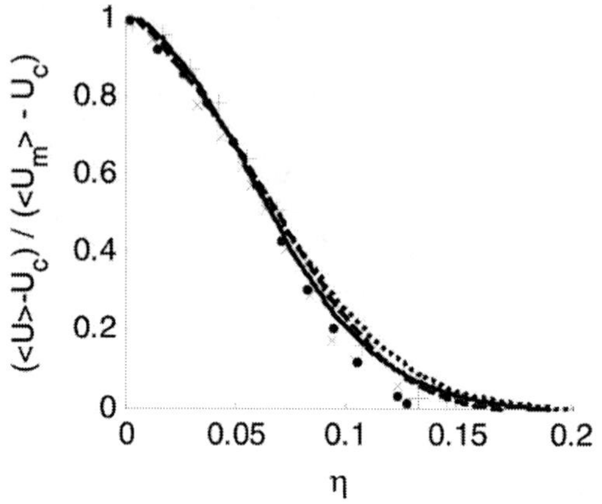

Figure 4. Mean streamwise velocity distribution across the jet. Experimental data [14] (symbols): + x/D = 70; • x/D = 80; × x/D = 90. Simulation results (lines): Solid 0.8M mesh; dashed: 1.2M; dotted 1.6M.

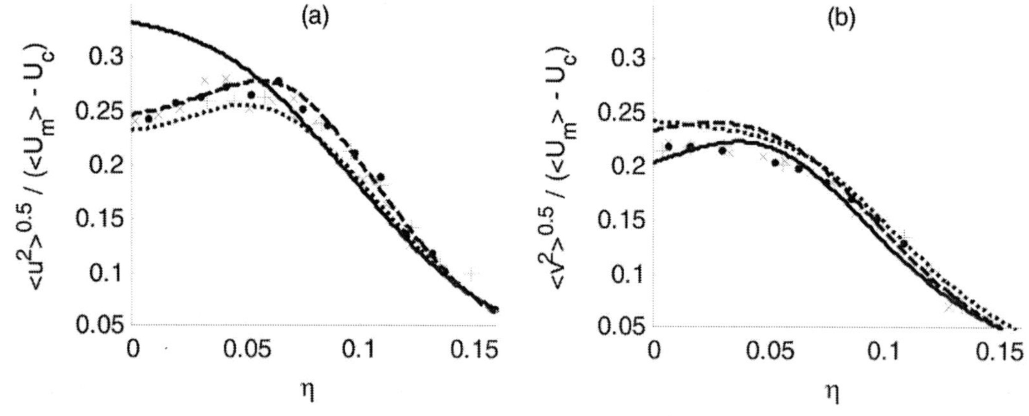

Figure 5. Streamwise (a) and radial (b) velocity fluctuation distributions across the jet. (see Figure 4 for legend).

Comparison of SGS scalar flux models

Once the proper mesh size was selected and the agreement of the simulation with experimental data was corroborated, the ability of LES to predict the scalar field using different closures strategies for modeling the unknown SGS scalar flux term in Eq (3) was evaluated. The three modeling strategies were: eddy diffusivity model with constant $Sc_T = 0.62$ (labeled as "Eddy" in the figures), eddy diffusivity model where the Sc_T is dynamically calculated (labeled as "Dynamic") and the dynamic anisotropy model (labeled as "anisotropy"). Figure 6 compares the simulated radial mean and fluctuation distribution of the scalar concentration (normalized by the centerline value) with the experimental data [14].

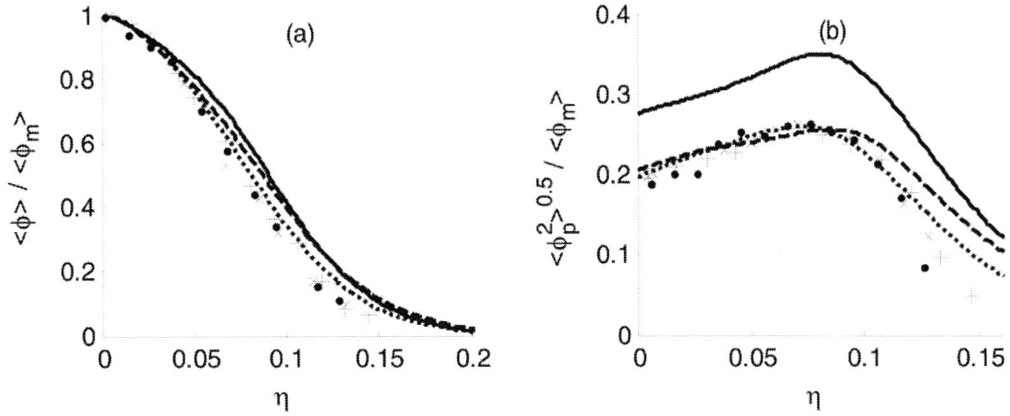

Figure 6. Radial mean (a) and fluctuation (b) concentration distribution across the jet.. Experimental data [14] (symbols): + x/D = 70; • x/D = 80; × x/D = 90. Simulation results (lines): Solid: Eddy; dashed: dynamic; dotted: Anisotropy.

Although the LES predictions of the radial concentration distribution are similar for the tested SGS scalar flux models in Figure 6, the dynamic anisotropy model provides a better performance. At low values of η, the three models agree with the experimental data. However, as the distance from the centerline increases, both for the mean (Figure 6a) and fluctuations (Figure 6b) the performance of the dynamic anisotropy model is much better than the other two. The use of a constant Sc_T number seems, therefore, not correct for the liquid-phase flow considered in this study.

The dynamic and anisotropic models tend to overestimate the scalar distribution for $\eta > 0.11$. The reason for this disagreement may be that the lateral mesh resolution in this region is not fine enough, which can explain the departure from experiments. However, the effect of the non-resolved flow structures (modeled by the SGS scalar flow models) on the mixing process are better accounted for by the anisotropy model as it shows the best agreement with the experimental data. The computational costs of the dynamic and anisotropy models were 1.2 and 1.7 times that of the eddy diffusivity model, respectively. When a higher level model has to be considered, it is usual that more computational resource is required.

Structure of the Flow: Qualitative Description

A highly unsteady behavior of velocity and concentration could be observed from the animations of the LES results. Instantaneous concentration isosurfaces of rhodamine B are shown in Figure 7a, where a cut has been made at z=0, in order to have a better perspective of the internal part of the jet. The isoconcentration lines of this plane are projected at the plane z=31.5 mm. The contour plot in the plane xz, y=0 is also shown in Figure 7.

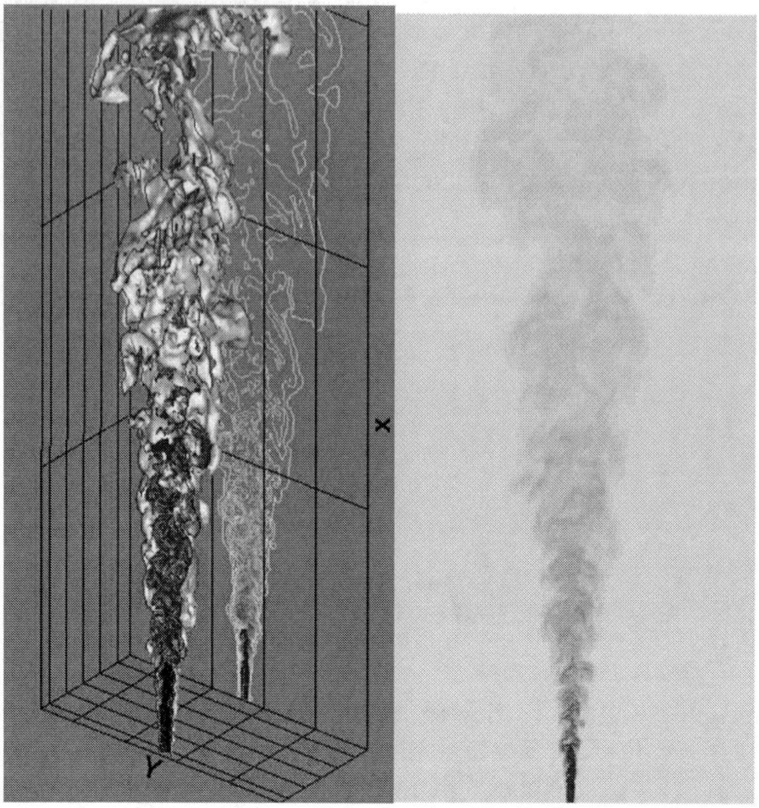

Figure 7. Instantaneous concentration of rhodamine B. Left: Isosurface plot (cut at z=0); Right: contour plot (plane xz, y=0). Darkest denotes maximum concentration.

In Figure 7 at the orifice exit the high-speed flow yields a laminar shear layer with smooth instabilities. However, this shear layer is unstable and propagates fast. Instabilities of the Kelvin-Helmholtz type enhance vortices roll-up; particularly where the inner potential core is almost extinguished. Vortices entrain jet fluid into the coflow stream and vice-versa as is shown in Figure 8.

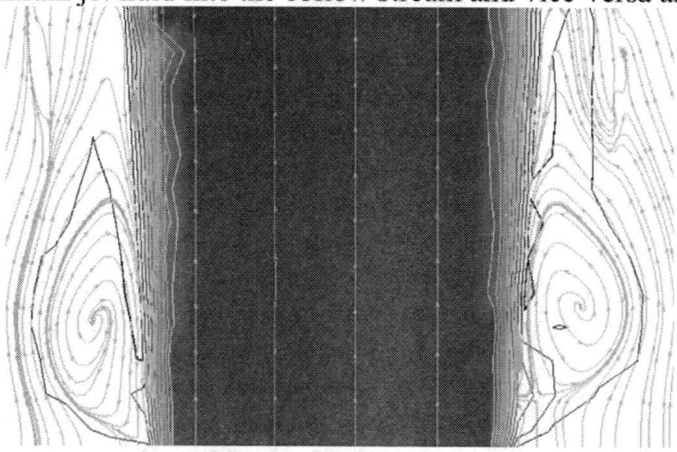

Figure 8. Instantaneous concentration of rhodamine B and velocity streamlines at the tube exit (plane xz, y=0). The color map of the Iso-concentration lines was been inverted to improve visualization.

In Figure 8 a vortex structure and two small counter-acting vortices are observed at each side of the plane. The lines of low concentration of the passive scalar in the coflowing stream orginate from the entrainment process of the ring vortex structure. Vortical structures develop further shear layers in the intermediate region, increasing gradually the mixing process (see Figure 7).

Summary and concluding remarks

In this work, LES of a turbulent liquid jet was carried out. A Smagorinsky model (with a dynamically calculated coefficient) was used to represent the SGS stresses. The performance of three SGS scalar flux models: eddy diffusivity model with both constant and dynamically-calculated turbulent Schmidt number (Sc_T) and the dynamic anisotropy model for closing the filtered scalar balance equation was evaluated. A mesh size sensitivity analysis showed that, although results from a 1.6M mesh were closer to experimental data, a 1.2M mesh provides very similar results with a lower computational cost. LES predicted the self-similar behavior observed for this flow. Mean and fluctuation velocity distributions were well captured by the numerical simulations; however, this was not the case for the virtual origin and the jet decay factor which might be influenced by mesh resolution and inflow conditions.

The linear eddy model with dynamic calculation of the Sc_T and the anisotropy model provided good agreement with experiments for the mean and the fluctuations of the passive scalar field. The anisotropy model gave a better approximation when capturing the non – linear interactions of the turbulent scalar fluxes. However, its computational cost was 1.7 times higher than the linear eddy diffusivity model. Consequently, LES, along with the anisotropy model for closing the filtered scalar transport equation, is more suitable for predicting, analyzing and optimizing processes taking place in liquid phase.

Acknowledgments

Authors thank the DAAD-COLCIENCIAS under collaboration PROCOL 2007 for a traveling grant. The first author acknowledges especially the Institute of Energy and Power Plant Technology (EKT) at the Technische Universität Darmstadt (Germany) where the main part of this work was conducted. Authors thank also the Faculty of Mines (U.N.C.) for traveling support.

References

1. Pope, S.B. (2000). *Turbulent flows*. Cambridge University Press, Cambridge.
2. Michioka T, Komori S. (2004). "Large-Eddy Simulation of a Turbulent Reacting Liquid Flow". *AIChE Journal*, 50 (11), pp. 2705-2720.
3. Hartmann H, Derksen JJ, van den Akker HEA. (2006). "Mixing Times in a Turbulent Stirred Tank by Means of LES". *AIChE Journal*, 52 (11), pp. 3696–3706.
4. Onishi R, Komori S. (2006). "Thermally Stratified Liquid Turbulence with a Chemical Reaction. *AIChE Journal*, 52 (2), pp. 456-468.
5. Revstedt J, Fuchs L, Kovács T, Trägårdh C. (2000). "Influence of Impeller Type on the Flow Structure in a Stirred Reactor". *AIChE Journal,* 46 (12), pp. 2373-2382.
6. Roussinova V, Kresta SM, Weetman R. (2003). "Low frequency macroinstabilities in a stirred tank: scale-up and prediction based on large eddy simulations". *Chemical Engineering Science*, 58, pp. 2297–2311.

7. Eidson T. (1985). "Numerical simulation of the turbulent Rayleigh-Bénard problem using subgrid modeling". *Journal of Fluid Mechanics*, 158, pp. 245-268.
8. He GY, Hsu T. (1999). "The effect of schmidt number on turbulent scalar mixing in a jet-in-crossflow". *International Journal of Heat and Mass Transfer*, 42, pp. 3727-3738.
9. Gallerano F, Napoli E. (2000). "Balance equation of the generalised sub-grid scale (SGS) turbulent kinetic energy in a new tensorial dynamic mixed SGS model" , *Continuum Mechanics and Thermodynamics*, 12, pp. 79-94.
10. Horiuti K. (1997). "A new dynamic two-parameter mixed model for large-eddy simulation". P*hysics of Fluids*.; 9(11), pp. 3443-3464
11. Löffler M, Pfadler S, Beyrau F, Leipertz A, Dinkelacker F, Huai Y, Sadiki A. (2008). "Experimental determination of the Sub-grid Scale Scalar Flux in a Non-Reacting Jet Flow". *Flow, Turbulence and Combustion*, 81 (1-2), pp. 205-219.
12. Huai Y, Sadiki A. (2007) . "Large Eddy Simulation of Mixing Processes in Turbulent Liquid Flows with Chemical Reactions", *Proc. of the 5th Int. Symp. on Turbulence and Shear Flow Phenomena*, Germany (Munich), pp. 1137-1142,
13. Huai Y. (2005). "Large eddy simulation in the scalar field". *Doctoral thesis*.. Technische Universität Darmstadt. Germany.
14. Antoine Y, Lemoine F, Lebouché M. (2001). "Turbulent transport of a passive scalar in a round jet discharging into a co-flowing stream". *European Journal of Mechanics: B/Fluids*, 20, pp 275-301.
15. Smagorinsky J.S. (1963) "General circulation experiments with the primitive equations I. The basic experiment. *Monthly Weather Review*, 91 (3), pp 99-164.
16. Germano M, Piomelli U, Moin P, Cabot W.H. (1991). A dynamic subgrid-scale eddy viscosity model". *Physics of Fluids*, 3(7), pp. 1760-1765.
17. Lilly D. (1992). "A proposed modification of the Germano subgrid-scale closure method", *Physics of Fluids*, 4 (3), pp. 633-635.
18. Pantangi P, Huai Y, Sadiki A. (2010) "Mixing Analysis and Optimization in jet mixer systems by means of large eddy simulation". In: Bockhorn H, Mewes D, Peukert W, Warnecke HJ. Micro *and Macro Mixing, Analysis, Simulation and Numerical Calculation*, series: Heat and Mass Transfer.
19. Durst F, and Schäfer M. (1996). "A parallel blockstructured multigrid method for the prediction of incompressible flow". *International Journal for Numerical Methods in Fluids*, 22, pp. 549–565
20. Mi J, Nobes DS, Nathan GJ. (2001) "Influence of jet exit conditions on the passive scalar field of an axisymmetric free jet". *Journal of Fluid Mechanics*, 432, pp. 91-125.

(522f) Scale-up of the Production of Monofunctional Polysiloxanes

Georg Witek[1], **Matthäus Siebenhofer**[2], Frank Uhlig[1] and Enes Aksamija[2], (1)Institute of Inorganic Chemistry, Graz University of Technology, Graz, Austria, (2)Department of Chemical Engineering and Environmental Technology, Graz University of Technology, Graz, Austria

Monofunctional PDMS are interesting building blocks for various inorganic-organic hybrid materials such as block copolymers, core-shell-particles and similar compounds.[1,2] However, the production of these materials on an industrial scale is limited.

Main target of this project is the development of a continuously operated reactor system for the production of monohydrido-substituted polysiloxanes with a narrow molecular weight distribution. The catalyst-free ring opening polymerization (ROP) of hexamethylcyclotrisiloxane (D_3) with *tert*-butyllithium (t-BuLi) is a promising route for this process (Figure 1).

In a first step, the kinetics and thermodynamics of the individual reactions - initiation, propagation and termination - were investigated. The ROP is highly selective towards the monomer D_3. This substance is selectively polymerized in the presence of other cyclic oligosiloxanes, such as octamethylcyclotetrasiloxane (D_4) or higher analogs.

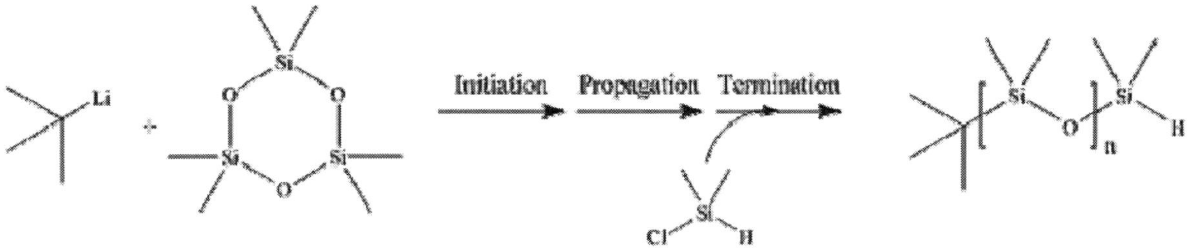

Figure 1: Ring opening polymerization (ROP) of hexamethylcyclotrisiloxane (D_3) with *tert*-butyllithium (t-BuLi) and endcapping with chlorodimethylsilane (CDMS).

In a second step a lab-scale plug flow reactor (PFR) was successfully operated. Priliminary experiments in a stainless steel pipe with static mixers showed a higher degree of polymerization than expected. An interaction of the polymerization reaction with the braze joint at the reactor inlet is assumed.

To solve this problem a PTFE plug flow reactor was used. Because of laminary flow conditions the tubing was coiled on a steel pipe to create radial mixing due to Dean vortices. To provide isothermal polymerization the reactor setup was placed in a temperature controlled water bath. With this experimental setup monofunctional PDMS with desired chain lengths and narrow molecular weight distributions (PDI ≤ 1,2) were accessible with high reproducibility.

[1] P. J. Miller, K. Matyjaszewski, *Macromolecules* **1999**, *32(26)*, 8760-8767

[2] J. Pfeiffer *5th European Silicon Days* **22.09.09**, Vienna/Austria, *Oral Presentation*

AUTHOR INDEX

AUTHOR INDEX